Mitchell G. Williams, Editor

# LAND SURVEYS

## A Guide for Lawyers and Other Professionals

**THIRD EDITION**



SECTION OF REAL PROPERTY | TRUST & ESTATE LAW

Defending Liberty
Pursuing Justice

Cover by Bobbie Sanchez/ABA Publishing.



Printed in the United States of America

16 15 14 13 12     5 4 3 2 1

**Library of Congress Cataloging-in-Publication Data**

Land surveys / Edited by Mitchell G. Williams.—3rd ed.
p. cm.
Includes bibliographical references and index.
ISBN 978-1-61438-652-0 (print : alk. paper)
1. Surveying—Law and legislation—United States. 2. Surveyors—Legal status, laws, etc.—United States. I. Williams, Mitchell G.
KF683.A75L36 2012
346.7304'32—dc23

2012026034

# Contents

# Biographies

## The Editor

***Mitchell G. Williams*** received a BA from the State University of New York at Binghamton in 1973, an MA from the State University of New York in Binghamton in 1975, and graduated summa cum laude from New York Law School in 1984 where he was Articles Editor of the *New York Law School Law Review.* He is a partner in the Real Estate Practice Group at Hunton & Williams LLP in New York and a New York State-licensed professional land surveyor. Mr. Williams is a member of the American Bar Association, the American Congress of Surveying and Mapping (ACSM), the National Society of Professional Surveyors (NSPS), and the New York State Bar Association (NYSBA). He has published a number of articles on law and surveying in PROBATE & PROPERTY, SURVEYING AND MAPPING, and THE JOURNAL OF LEGAL HISTORY, and was the Editor of LAND SURVEYS, A GUIDE FOR LAWYERS AND OTHER PROFESSIONALS (2nd ed. 2000).

## The Authors

***Richard F. Bales*** is employed by Chicago Title Insurance Company as Assistant Regional Counsel and Assistant Vice President in its Wheaton office. He graduated from Illinois College in Jacksonville in 1973. He received his juris doctorate degree from Northern Illinois University in 1983 and is admitted to practice law in Illinois. As a member of the Education Committee of the Illinois Land Title

Association, he has lectured throughout the state on all aspects of title insurance and real estate law. He has also written numerous articles for the Chicago Bar Association, the DuPage County Bar Association, and the Illinois State Bar Association. These articles include *Land Surveyor Liability to Third Parties in Illinois*, which appeared in the March 2007 issue of the ILLINOIS BAR JOURNAL, the peer-reviewed journal of the Illinois State Bar Association. In 1999, several of his articles appeared in a book on survey issues that was published by the American Bar Association. Mr. Bales's book, A GUIDE TO RESIDENTIAL AND COMMERCIAL SURVEYS IN ILLINOIS, was originally published by the Illinois Institute for Continuing Legal Education in 2004 and has since then been updated and reissued in 2011. Mr. Bales is a member of an American Land Title Association and American Congress on Surveying and Mapping liaison committee, and he helped draft both the 2005 and 2011 ALTA/ACSM land title survey standards. These standards are used by surveyors throughout the United States. Mr. Bales's article on the 2011 survey standards was published by the American Bar Association in the July/August 2011 issue of PROBATE & PROPERTY. Mr. Bales is also the author of THE GREAT CHICAGO FIRE AND THE MYTH OF MRS. O'LEARY'S COW, a book wherein he claims (through the study of pre-Chicago Fire land records) to have solved the mystery of who started the Great Chicago Fire of 1871. His research was the subject of a Discovery Channel "Unsolved History" episode and an episode of the Weather Channel's television show, "When Weather Changed History." Mr. Bales is now working on a biography of George Wellington "Cap" Streeter and the turbulent post-Chicago Fire saga of Chicago's "Streeterville" neighborhood.

***Landon Blake*** is licensed to practice land surveying in California and Nevada. He graduated with an AAS from Flathead Valley Community College in Kalispell, Montana. He has been employed for the past ten years as a project manager for a small civil engineering and land surveying company in California. His work typically involves surveying for right-of-way acquisition and construction of large flood control and transportation infrastructure projects. Mr. Blake is a charter member of the OSGeo, the volunteer editor of the OSGEO JOURNAL, and a member of the California Land Surveyor's Association Central Valley Chapter. He is a volunteer programmer of the open source desktop GIS program OpenJUMP and the founder of the California Central Valley Geospatial Professionals Group. He writes a regular column on boundary surveying for THE AMERICAN SURVEYOR MAGAZINE.

***Robert W. Foster*** is a Registered Professional Surveyor and a Registered Professional Engineer (Civil) with over fifty years' experience, having graduated from the University of Vermont in 1955 with a BS in Civil Engineering. Mr. Foster has specialized in offering professional consulting services in arbitration, dispute resolution, and litigation involving surveying and civil engineering issues. Although semi-retired, he is an active member of the Dispute Resolution Panel of Neutrals of REBA Dispute Resolution, Inc., a subsidiary of the Real Estate Bar Association of Massachusetts. Mr. Foster is a past president of the American Congress on Surveying and Mapping and a past president of the International Federation of Surveyors, a federation of over one hundred national professional surveying associations. He has provided professional testimony in litigation involving property disputes, appeals of permit denials, eminent domain proceedings, and professional negligence, and has testified before the United States Congress and the Massachusetts legislature on pending legislation and budgetary matters.

***Gerald J. Greenan*** is past chair of the 250-member Real Estate Committee of the Erie County Bar Association and a frequent lecturer at continuing legal education seminars. He has focused his practice on Real Property Law since his admission to the New York Bar in 1962. In the region where he practices, urban, suburban, and rural western New York, the majority of real estate transactions close without fee title insurance and are based upon an attorney's opinion of the state and quality of title. Most of his clients are attorneys who rely upon his examination of the record title and his written opinion of the contract marketability of their clients' purchases.

***Knud E. Hermansen*** is an attorney, professional engineer, and professional land surveyor licensed in several states. His education includes a BS and PhD in Civil Engineering from Pennsylvania State University, an MS in Civil Engineering (emphasis on surveying and mapping) from the University of Wisconsin, and a JD from West Virginia University. Mr. Hermansen is a professor at the University of Maine and operates a consulting firm offering surveying, engineering, and legal services. His professional memberships include the American Congress on Surveying and Mapping, American Society of Civil Engineers, National Lawyers Association, and several state survey societies and bar associations. He is an author or coauthor of the following titles: *Maine Roads and Easements*; *Writing Legal Descriptions*; *Boundary Principles and Procedures for Pennsylvania*; *A Partial Compendium of Rules and Regulations Pertaining to Single Family Residential Development in Maine*; *Legal Topics in Boundary Surveying*; *An Introduction to Surveying Management*; and *Handbook of Annotated Forms for the Surveying Practice*.

***R. Lee Hixson*** began his surveying career by putting himself through college working on the survey crew of the Utah Power & Light Company. After graduating from the University of Utah with a BS and MS in sociology he moved to northern California. Two years of teaching at a junior college led him to return to surveying, and he became a Professional Licensed Surveyor in California in 1980. Mr. Hixson is now licensed in six additional western states: Oregon, Washington, Idaho, Wyoming, Utah, and Nevada. He specializes in research, analysis, and boundary resolutions relating to a variety of mapping projects. He has completed fifty or more ALTA/ACSM Land Title Surveys and recorded over two hundred maps in three states, most of them in California. He has been published in all of the major nationwide journals, with most of the articles focusing on the need for a higher standard of boundary resolutions.

***Gary R. Kent*** is chair of the National Society of Professional Surveyors' committee on the ALTA/ACSM Standards and is liaison to NSPS for the American Land Title Association. He is in his twenty-eighth year with The Schneider Corporation, a surveying, GIS, and consulting engineering firm based in Indianapolis and with offices in Charlotte and Des Moines. Mr. Kent is a past president of both the American Congress on Surveying and Mapping and the Indiana Society of Professional Land Surveyors. He taught Boundary Law, Legal Descriptions, Property Surveying, and Land Survey Systems as an adjunct instructor for Purdue University in Indianapolis and West Lafayette from 1999 to 2006 and received Excellence in Teaching and Outstanding Associate Faculty awards for his efforts there. Gary is currently in his third term as vice chairman of the Indiana State Board of Registration for Land Surveyors. He also writes regular columns for THE AMERICAN SURVEYOR magazine and for the ACSM BULLETIN.

***Wendy Lathrop*** is a self-employed consultant, licensed as a Professional Land Surveyor in New Jersey, Pennsylvania, Delaware, and Maryland, and as a Professional Planner in New Jersey. She is also a Certified Floodplain Manager (CFM) through the Association of State Flood Plain Managers. Since 1974, her surveying experiences have covered a broad range of projects in the field and in the office. She is president of Cadastral Consulting, LLC, which provides consulting services and continuing education courses for surveyors. Ms. Lathrop is a contributing editor for THE AMERICAN SURVEYOR magazine.

***Harlan J. Onsrud*** is a professor in the School of Computing and Information Science at the University of Maine where he teaches graduate courses in information systems law and responsible conduct of research. As an attorney, engineer, and land surveyor, his research has focused on the interface of these fields and particularly on analysis of legal, ethical, and societal issues affecting the creation and use of digital databases and assessment of the effects of location tracking technologies. He received a BS and MS from the University of Wisconsin College of Engineering and a JD from the University of Wisconsin Law School.

***David E. Snediker*** is a partner in the firm of Paul Hastings LLP, resident in its New York City office. He focuses his law practice on commercial real estate matters, concentrating in the areas of pension fund real estate activities and real estate finance matters. Mr. Snediker is a contributing author to THE COMMERCIAL REAL ESTATE TENANT'S HANDBOOK (1987) and to MORTGAGE FORECLOSURES IN NEW YORK

(1996). He has lectured as a faculty member at numerous NACORE, AECRE, ABA, and local bar association functions on topics relating to commercial real estate matters. Mr. Snediker has been a member of the American Bar Association's Committees on Creditors' Rights in Real Estate Financing and Land Surveys. He is the firm's voting member for the Pension Real Estate Association, and he is a former member of PREA's Governmental Affairs Committee. He is a former chairman of the board of directors of the International Association of Attorneys and Executives in Corporate Real Estate, and he is currently a member of the New York State Bar Association's Committee on Title and Transfer and Real Estate Finance. He is also an independent director of Fieldpoint Private Bank and Trust, where he is on the Loan Committee and the ALCO Committee and is chairman of the bank's Compensation Committee. Mr. Snediker received his BS in 1969 from the University of Notre Dame and his JD in 1975 from St. John's University School of Law.

***Donald A. Wilson*** is president of Land & Boundary Consultants, Inc., based in southeastern New Hampshire, and has been in practice as a land boundary consultant for over forty years. While most of his practice has been in and around the New England area, Mr. Wilson has served as a consultant to firms and groups throughout the United States and Canada. He is both a licensed land surveyor and professional forester, and has conducted programs on a variety of topics, including description interpretation, boundary evidence, law, title problems, forensic procedures, and related subjects. Mr. Wilson has been active in numerous organizations in the surveying and mapping, forestry, and related fields, including the American Congress on Surveying and Mapping, the Canadian Institute of Geomatics, the American Society of Photogrammetry and Remote Sensing. He is a life member of the American Forestry Association, a Fellow of ACSM, past president of both the Maine Society of Land Surveyors and the New Hampshire Land Surveyors Association, and a member of the New Hampshire Title Association and the New England Land Title Association. In addition to more than two hundred technical publications in several areas, Mr. Wilson has been involved with over fifty books. He is coauthor of Evidence and Procedures for Boundary Location and Boundary Control and Legal Principles and author of Deed Descriptions I Have Known But Could Have Done Without, Easements and Reversions, and the chapter on Boundary Surveys in the Surveying Handbook. His two latest works, published through John Wiley & Sons, are Interpreting Land Records and Forensic Procedures for Boundary and Title Investigation. Mr. Wilson has spoken before and consulted with numerous organizations, including several state Departments of Transportation, several power authorities, offices of Attorneys General, the International Right of Way Association as well as Minatome Corporation, the U.S. Forest Service, and the Bureau of Land Management. Mr. Wilson formerly contributed a regular column for Professional Surveyor magazine and is presently an instructor for RedVector in producing online professional courses. He is also part owner of and lead instructor in Surveyors Educational Seminars and a regular seminar coordinator for the University of New Hampshire.

# Preface

It has been ten years since the second edition of this book appeared. A lot has changed, and much has remained the same.

The Minimum Standard Detail Requirements for ALTA/ACSM Land Title Surveys have been revised twice, first in 2005 and more extensively in 2011. The ALTA forms of title insurance policies were extensively revised in 2006. There have been significant changes in surveying technology, particularly in the Global Positioning System which allows greater precision in measurement at a lower cost.

But people still fight over the location of their property lines and encroachments, and existing parcels continue to be subdivided into smaller lots and legal descriptions have to be written. Lawyers still hear their clients complain about getting a new survey: after all, the building didn't move, I have title insurance, and so forth. But as you will see from the articles in this book, new surveys still matter in sales and purchases of real property and in the financing of those sales and purchases.

Thirteen of the articles that follow appeared in the second edition, but nearly all of them have been extensively revised and updated to reflect the changes in the last ten years. There are also thirteen new articles: they include an analysis of the new 2011 ALTA/ ASCM minimum survey standards coauthored by the chairman of the Liaison Committee that drafted them, the role of aerial photography in surveying, the National Flood Insurance Program, and "how to" articles on reviewing survey maps.

PART I

# *Land Surveys: An Introduction*

# What Every Lawyer Should Know About Title Surveys

1

Mitchell G. Williams
Harlan J. Onsrud

A complete and accurate land survey is of fundamental importance in nearly all real estate transfers. A comprehensive land survey and physical inspection of the property is the only efficient and reliable means of delineating the physical limits of the property and locating the improvements on it. Yet land surveys are one of the least understood and most frequently overlooked elements in a real estate transaction. This article outlines the basic reasons for obtaining an accurate survey, discusses various types of land surveys, presents an example of how a surveyor might evaluate field and record evidence in accomplishing a property line survey, and then concludes with some suggestions for lawyers and title insurers to assure that an adequate survey is obtained.

## I. What Is a Survey?

The word *survey* is derived from an old French word meaning "to look over" and refers to the process of evaluating real property evidence in order to locate the physical limits of a particular parcel of land. The real property evidence considered by the surveyor typically consists of physical field evidence, written record evidence, and field measurements. The surveyor, having made an evaluation of the evidence, forms an opinion as to where he believes a court would locate the boundary lines of the property. The typical modern-day surveyor sees himself as an expert evaluator of evidence and would expect to arrive at the same opinion of boundary location regardless of who commissioned the survey. The surveyor's opinion is founded on experience and applicable legal precedents; unlike an attorney, the surveyor does not see himself primarily as an advocate for his client.

The surveyor then prepares a "map or plot of survey" to communicate his opinion to others. The map or plot is colloquially known as the survey. It is important to remember, however, that the map and accompanying notes represent a report of the survey, with the survey itself being primarily a measurement and evidence evaluation process.

While land surveying is often associated with engineering, the two professions are distinct. The evaluation of land surveying evidence is not a "science" in the sense that there is one procedure to follow which will yield the "correct" result. Surveyors occasionally disagree on the proper location of a boundary line—not necessarily because one surveyor measures better than the other, but more commonly because each surveyor has weighed the evidence differently and they have formed different opinions. Just as two lawyers may draw different conclusions from the same line of cases, surveyors may disagree about the appropriate location for a boundary. Because a survey is a professional opinion, it is subject to review by a court in the event that a boundary dispute reaches litigation. For the same reason, attorneys should remember that a survey and supporting documentation provided by one professional surveyor may be far superior or far inferior to that provided by another. Almost any field technician with basic training can make measurements with an acceptable degree of precision and replicability. In those instances in which locating the bounds of a land parcel requires an extensive amount of evidence evaluation, an individual with the requisite amount of education, knowledge, and experience should be employed to accomplish the survey with competence.

## II. Why a Survey?

There are six fundamental reasons for requiring land surveys in real estate transactions:

1. *The Existence of the Property.* Nearly all titles to land in the United States depend on an original grant or patent and subsequent conveyance instruments. Each of these instruments contains descriptions of the land conveyed. It is a fundamental principle that for a deed to be valid it must contain a sufficient description. Whether a metes-and-bounds description or a description by reference to a parcel on a map is sufficient to transfer the property often depends upon whether a knowledgeable surveyor can interpret the description to reasonably locate the property physically on the ground. In determining whether the land description is sufficient, the surveyor determines whether the description forms a mathematically

closed figure and whether the description reasonably conforms to the physical evidence on the earth's surface. The first determination is done by numeric calculation, the second by physical measurements in the field.

2. *The Relationship of the Property to Adjoining Properties.* Merely locating the lines described in a deed on the ground is not adequate to establish the physical limits of a property owner's interest. All parcels of land exist in relation to the parcels surrounding them. Surrounding parcels may include privately or publicly owned lands, rights-of-way, easements, roads, streams, and other bodies of water. At some point in the past, all adjoining land parcels were held in common by a single grantor. Over time, parcels were partitioned off or subdivided to arrive at the current ownership configuration. As a general rule, the description in a senior deed or prior conveyance controls over any discrepancy in a later one. If the drafter made an error or created an ambiguity in describing a parcel being partitioned off from a larger parcel, or made an error in a later attempt to "correct" or refine an earlier description, the legal descriptions of adjoining parcels may be inconsistent. The two descriptions of their "common" boundary may in fact either overlap or have a gap between them. Failure to discover overlaps may leave the holder of the junior deed owning much less property than the junior deed on its face would indicate. The presence of gaps or gores also poses problems when attempting to consolidate several adjacent parcels under a single owner for development purposes. When consolidation is attempted, one must definitively establish ownership to these leftover land strips. If a gap or gore exists along a street line or right-of-way, it has the potential of creating a landlocked parcel.
3. *The Relationship of Occupied Lines to Record Lines.* Not infrequently, the boundary lines of a parcel as physically occupied or possessed by its owner differ from the distances and direction or the monuments called for in the deed. Discrepancies between possession and the record deed lines may range from minor variations in fence line locations to substantial encroachments of multistory buildings. A land survey should always show the occupied lines together with the deed record lines and the extent of any mismatch. Significant mismatches may suggest potential claims of ownership by senior right or adverse possession or a change in a boundary line by mutual agreement and acquiescence. To cut off any potential rights of another to a claim of adverse possession, the property owner may want to record an appropriate document confirming his claim of ownership or seek a change in possession to match the record lines.
4. *The Location of Physical Improvements.* This reason for requiring a survey is related to the previous one, but deals with the relationship of all physical improvements on the parcel to the boundary lines of the parcel, not just those improvements near the exterior limits of the parcel. Features that surveyors are often requested to locate include fences, walls, driveways,

pavements, buildings, structures, utilities, wells, and natural features such as streams and ponds. This information is necessary to determine the presence of features that may limit the value or use of the property, and to determine conformity to setback lines contained in recorded documents and with local zoning ordinances regarding minimum building setbacks. It is also necessary to confirm that the improvements do not encroach upon easements or rights-of-way. When most attorneys and laypersons think of a survey, this is the type of information they expect to see on the surveyor's final survey map.

5. *Unrecorded Easements and Other Facts Not of Record.* There are numerous unrecorded rights that can affect title to land which may not show up in a title search but will become obvious upon an inspection of the property. The right of a neighbor to use utility lines, drainage ditches, sewer lines, and unrecorded travel easements across the property may have arisen by prescription or other methods of unwritten land transfer. A visual inspection of the property will usually give some physical indication as to whether such adverse rights may exist; for example, the presence of manholes or vent pipes suggests underground sewers or other utilities. Typically, only a survey in which unrecorded physical features are referenced to the property lines will induce a title insurance company to remove its exception in regard to "any state of facts an accurate survey might show."
6. *Water Boundaries.* Water boundaries are a complex subject, and details of ownership, usage rights, and location vary from state to state. These issues can only be dealt with briefly in an article of this scope. In general, a boundary defined by a body of water—for example, a creek, river, bay, or oceanfront—moves with changes in the course of the creek, river, or high-water mark over time, provided that the change is gradual; changes due to sudden causes, such as severe flooding or hurricanes, or changes due to actions by the property owner, whether gradual or not, such as building a dam, dredging, or filling, do *not* result in the movement of the boundary. It can be difficult to determine what caused a change in the course of a stream or in the high-water mark without historical research into old maps, deeds, and surveys. If the water in question is "navigable," a number of additional issues arise. A body of water is navigable if it in fact can physically be used for commerce or navigation, e.g., a river or a lake connected to other navigable waters, or if it is subject to the ebb and flow of the tide, e.g., oceanfront or bayfront property and tidelands. The title to land underlying waters that were navigable at the time of a state's admission to the Union was reserved to the state. The federal government, however, under the commerce clause, retained a "navigable servitude" over all waters that are presently, have been in the past, or may be in the future susceptible to use for purposes of interstate or foreign commerce. As a practical matter, this means that any portion of a body of water that was

navigable when a state was admitted to the Union and which has since been filled in, whether or not the fill was done with the permission of, or a license from, the Army Corps of Engineers or other governmental authority, is still "navigable" and subject to the federal navigable servitude.[1] The federal government can require that the fill, together with any improvements, be removed if it is necessary to improve navigation. This is not a taking or an exercise of eminent domain, and no compensation is required to be paid. This can be a major concern for marinas and other waterfront properties, many of which have significant improvements on filled-in land. In such circumstances, it would be prudent for a purchaser or lender to have the surveyor show, to the extent determinable, the location of the high-water mark at the time the applicable state entered into the Union, as well as any evidence of any filled-in lands. A title policy issued for such properties will contain an exception from coverage for "the rights of the federal government to enter upon and take possession without compensation of lands now or formerly lying below the high-water mark."[2]

## III. Types of Surveys

The above six reasons for obtaining a survey are fundamental in the case of a title survey. There are, however, many types of land surveys: boundary surveys, title surveys, topographic surveys, plot plans, subdivision maps, "as-built" surveys, and so forth. Each serves a different purpose, and they are not interchangeable. Typically, a boundary survey will only show the record boundaries and improvements and evidence of possession that are close to the actual record lines; a title survey will typically include all significant improvements and features on the property as described above. A subdivision map will show existing boundary lines of record and proposed lines for new parcels; it would not show any improvements or evidence of possession at all. Plot plans and site plans typically show record lines and the proposed location of new construction and the approximate locations of existing improvements to help an architect or engineer locate and design the proposed construction. A topographic survey usually indicates the contour, shape, elevations, and physical features of the ground within certain boundaries; like the plot plan, it is designed to aid an architect or engineer in the design and layout of a building, not to give a professional opinion on the location of the boundaries. Some attorneys believe that if they acquire an "as-built" survey, they have acquired the highest quality survey available. An "as-built" survey,

1. It requires an act of Congress to release a body of water from the federal navigable servitude.

2. A "navigable servitude" endorsement is available which will insure against the forced removal of any such improvements as a result of the exercise of the navigable servitude. The issuance of that endorsement is an underwriting issue for the title insurers.

however, is merely a detailed map of a new building or other improvement and its relation, as built, to the *plans* from which it was built. It may appear complex and comprehensive to laypersons, but its preparation does not typically address boundary or title concerns. The purpose of an "as-built" survey is to determine if the completed project accords with previously approved plans and specifications. Absent some statement to the contrary, these types of survey do not address the location of the boundary lines, their relationship to buildings and improvements, or adjoining lots. An attorney who uses one of these types of survey to advise a client about boundary and title questions flirts with malpractice.

## IV. The Presumed Priority of Conflicting Title Elements that Determine Boundary Location

A surveyor looks to legal principles as a guide in evaluating the evidence for a boundary line location. One such principle is the presumed priority of conflicting title elements that determine boundary line location. A right of ownership can arise from two sources: by written means (such as a deed or written boundary line agreement) or by unwritten means (such as adverse possession). The resolution of conflicts between written and unwritten rights is one of the most difficult problems for both surveyors and lawyers. As the resolution is usually dependent on the particular facts in each case, it is beyond the scope of this article. But even within a deed or other written conveyance, there frequently are conflicting terms that the surveyor must resolve.

When such a conflict occurs, it is necessary to decide which terms were intended to control and which terms were informational. Which term is controlling generally depends upon its order of importance as determined by courts. One summarized listing of a judge-made priority ranking among the terms of a deed is as follows:

1. Call for a specific survey
2. Call for monuments
   a. Natural
   b. Artificial
   c. Record boundaries (in the event of a gap)
3. Calls for directions and distances
4. Calls for directions
5. Calls for distance
6. Call for coordinates
7. Call for area

If the results obtained by adhering to the above ranking are clearly contrary to the overall intent expressed in the deed, however, the intent expressed in the deed will control.

The above priorities are based upon assumptions about the relative certainty of each type of evidence. The presumption that directions should control when in conflict with distances is based on the assumption that one can determine angles and bearings more precisely than distances. In some jurisdictions, however, directions were historically observed with an imprecise magnetic compass and distances were measured with a precise steel tape; in these jurisdictions, courts held that distances should control over directions when in conflict. It makes sense that when the reasons for adhering to the presumed priority ranking no longer exist, the presumed ranking should fail and the best available evidence should prevail.

## V. Sample Evaluation of a Property Line Fact Situation

In order to illustrate the land surveyor's use of the foregoing rankings in arriving at an opinion as to where property ownership lines are likely to lie, we present the following hypothetical facts. Alice owns the large parcel of land shown in figure 1. The parcel was surveyed and corners monumented in 1932. There are no conflicts between Alice's lines and those of her neighbors. Alice sells a portion of her parcel to Betty in 1960. The land description in Betty's deed from Alice reads as follows:

> Beginning at the NW corner of Alice's property as marked by a 2" iron pipe; thence East along the centerline of White Road 400 feet to the center of the Jones Creek culvert; thence South 200 feet along Jones Creek; thence West 400 feet; thence North 200 feet to the point of beginning.

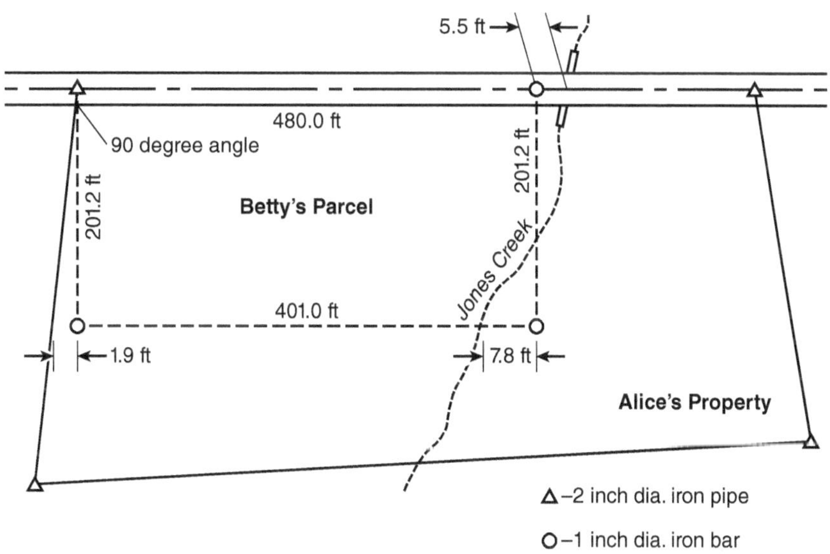


The information shown in figure 1 reflects the state of facts disclosed by a field survey. There are no fences, tree lines, hedges, or evidence of possession along the lines of Betty's parcel. There is no field or record evidence to suggest that Jones Creek has significantly changed its course since 1960.

How might a surveyor reasonably evaluate this evidence? One of the surveyor's obligations in accomplishing a land survey is to locate the title lines and then locate occupation lines with respect to those title lines. Under the given facts, it is proper to consult only the terms of Betty's deed description in locating the title lines.

- The call for the 2-inch iron pipe is a call for an artificial monument, and if the surveyor can establish that the iron pipe actually found is the pipe called for in the deed, it fixes the location of the point of beginning. The found 1-inch diameter iron bars shown in figure 1 are not called for in the deed and, therefore, are not legally artificial monuments. There is nothing to suggest that they were set during a survey, except that the bars are of a material and size typically used by surveyors in marking property corners and are in the approximate locations where a current survey would place Betty's property corners. These uncalled-for monuments should be accepted as marking the southerly line of Betty's parcel only if the lines of this parcel are so uncertain that most surveyors would not locate the title lines in the same locations.
- In the first course, the call "to the center of the Jones Creek culvert" is controlling as a natural monument called for in the deed. The 400-foot distance called for is informational. It provides an approximate distance to aid in finding the called-for physical corner.
- In the second course, the creek as a natural monument that is explicitly called for in the deed delineates the property line. Thus, the eastern line of Betty's parcel is synonymous with Jones Creek. If the called-for distance of 200 feet is accepted as controlling, it would be necessary to trace the sinuosities of the center of the creek to find the southeast corner of Betty's parcel. However, that distance may not necessarily be controlling when considering the overall intent of the deed. Further analysis is required.
- The call "thence West 400 feet" does not additionally say "to the east line of Alice's property." It appears unreasonable to assume, however, that Alice had intended to retain the small sliver of land that would result if the uncalled-for monument in the southwest corner is held as controlling. Additionally, when measured along Betty's southerly line, the creek is less than 400 feet away from Alice's west property line. Therefore Betty's parcel appears to be defined on its west end by Alice's west property line.
- The only side of Betty's parcel that has yet to be located is the southerly boundary. When considering the overall written intentions of the parties to the deed, it appears that the parties intended to partition off a parcel in the northwest portion of Alice's property bounded by Jones Creek on

the east, bounded by Alice's west property line on the west, and having dimensions of approximately 400 feet by 200 feet.

- The sequence of calls in a description is generally held by the courts to be immaterial. Therefore, reading the description backwards, one possible location for Betty's southwest corner is at a distance of 200 feet from the point of beginning along Alice's west property line. It would then appear reasonable to define the southerly line of Betty's parcel by a line extended from this monument and running easterly parallel to White Road. This would result in a configuration for the parcel in close agreement with the overall written intent of the parties.
- In the alternative, the court might find the southerly line of Betty's parcel to lie a perpendicular distance of 200 feet from and run parallel with the centerline of White Road. Because the grantor had the responsibility for making the description clear and failed to do so, any ambiguity in the description should be resolved in favor of the grantee. This configuration also closely agrees with the overall written intent of the parties. This is the solution that a large number of surveyors would probably reach.
- In some eastern states, particularly in rural areas, uncalled-for monuments tend to carry substantial weight in the courts. In such jurisdictions, the court would probably hold a straight line passing through the two 1-inch diameter bars as defining Betty's southerly property line. Due to the inconsistency in the dimensions between the monuments along the east and west bounds, however, it appears unlikely that a surveyor set the bars. They probably were set by a layperson to mark the approximate locations of the property corners and never intended to mark the precise actual locations of the corners. Under these facts, the 1-inch iron bars should not be held as marking the location of Betty's southerly line.

The primary purposes in presenting the preceding example are to illustrate the thought processes that a surveyor goes through in evaluating property line evidence and to stress that the location of property lines is more often a matter of legal opinion than scientific fact. The fact situation presented here is very simple in comparison to the fact situations actually encountered in practice. Surveyors can, and not infrequently do, come to different conclusions in evaluating property line evidence. It is important to reemphasize, however, that surveyors act primarily as expert witnesses and not as advocates. As an expert in evidence evaluation, the surveyor expects to arrive at the same conclusions from the evidence regardless of whether his client is Alice or Betty. The lawyer, on the other hand, represents exclusively his client's interests and advocates his client's side of the story to the very best of his abilities.

If a court upheld the surveyor's evaluation of the evidence in the example, it is because the surveyor arrived at a comprehensive and well-reasoned answer rather than because he arrived at the theoretically correct answer. Again, there are no "true" answers waiting to be discovered, only well-reasoned answers.

The preceding example has illustrated possible difficulties in locating the exterior bounds of a land parcel. The land surveying work product requested by the land owner or his attorney typically contains far more than a schematic diagram of the property lines and an opinion as to how the surveyor located the exterior bounds of the property. The more information requested from the surveyor, the more complex are the surveyor's tasks of research, measurement, legal evaluation, and reporting.

## VI. Survey Certifications

How is a lawyer or title company to know if a surveyor has actually considered all those matters that are of concern to the lawyer and whether the surveyor's conclusions represent a "well-reasoned" answer to the boundary location questions? All too often the survey delivered says no more than "Survey of . . ." or "Map of Property at . . . ," perhaps with a legend "certifying" to specific parties that an actual field survey has been accomplished. The lawyer should insist that the surveyor specify the code of practice or standards to which the survey was made. Customary standards of practice and written standards of practice vary considerably among the states and within each state. If the lawyer is providing advice in a transaction prior to the ordering of a survey, he should contractually obligate the surveyor to a particular desired set of standards and required certifications.

It is important for the lawyer to examine any code of practice used in order to determine if it adequately deals with the five major concerns listed in this article. If it does not, then the lawyer should ask the surveyor to address those concerns by carrying out any additional tasks required and adding specific certifications to the survey map. If a surveyor's contract did not obligate him to provide certain certifications, the surveyor may contend that he was not required to carry out the work that would allow him to make the requested certification. He justifiably may request further compensation in order to accomplish the additional work. In addition, there are many matters to which a surveyor should not be requested to certify. These include matters beyond his capable knowledge, such as whether underground encroachments of foundations exist. He also should not be requested to certify to matters that rightfully fall within the responsibility of the title insurance company, such as certifying that all easements have been shown on the survey regardless of whether disclosed by the title insurer or evident on the ground.

Codes of practice are most commonly promulgated by state surveying societies and by state licensing boards. In some states, there are codes of practice set forth in state law that a surveyor must follow. Within a state, there may be more than one professional society and, accordingly, multiple codes. The lawyer representing a client should be familiar with the code commonly used in the state. Some codes of practice deal purely with technical standards of precision in measurement and do not discuss what the survey should show, such as the Standards for Boundary Surveys in the State of Ohio (OHIO ADMIN. CODE ch. 4733-37). At the other end of the continuum are standards that treat at some length the issues of what research is required, what the survey should show, and what technical standards of measurement must be met, such as the joint ALTA/ACSM Minimum Standard Detail Requirements for Land Title Surveys, the most recent version of which was adopted in 2011 (the ALTA Code).

It has become common to ask for a "national" standard for land title surveys. Some have advocated the ALTA Code as a set of standards that could fulfill this need; however, an attorney should not blindly accept a survey certified in accordance with the ALTA Code. A client may require numerous items (e.g., flood zone, parking, and utility locations) that are not part of a standard ALTA survey but are listed in the ALTA standards as "optional," and must be specifically requested. Familiarity with the ALTA code and the client's needs are essential.

## VII. Conclusions and Recommendations

This article is intended to explain to the attorney why a proper and current land survey is important for the typical real estate transaction. This article has illustrated some of the intricacies of the land surveyor's reasoning process in the evaluation of boundary line evidence. Awareness of the complexity of a surveyor's task in accomplishing a land survey leads to the conclusion that, in contracting with a surveyor to carry out a survey, we should perhaps resort to published surveying standards to define the extent of the work to be accomplished. However, we have also illustrated that the lawyer must use published standards of practice with caution. Even state codes required by statute or the most widely accepted standards and codes of practice, such as the recent ALTA Code, will not meet the needs of all clients at all times.

We have suggested a few additional matters of survey that may concern your clients and that may be handled through the use of appropriate certifications. These are but a few of the many surveying matters attorneys must be aware of. To better address the concerns of their clients, attorneys must be aware of the purpose of each survey and must have an understanding of the proper limits of the surveyor's evidence evaluation, research, and measurement processes.

# What to Look for When Examining a Survey Map[1]

# 2

R. Lee Hixson

Most people are aware of the existence of malpractice in the legal and medical professions, and they have at least some clue about what steps can be taken to avoid becoming a victim of lawyers or doctors who do not follow accepted practice. But what about surveyors? Are lawyers working with survey maps even remotely aware of what constitutes surveying malpractice? When unfolding a map for the first time, do most lawyers have any idea what to look for to determine whether the boundary resolution shown on the map is a good one? Is it possible for a non-surveyor to spot a weak boundary on a map?

The answer to the last question is "yes." The purpose of this article is, first of all, to alert lawyers to the possibility that some maps are deficient and, second, to give them some guidance in differentiating between a map with a good boundary and one with a weak or poorly resolved boundary.

Boundaries are the single most important facet of any map produced by a surveyor. Most maps, if not all, show a boundary. Conditions vary from state to state, but, of the large variety of maps that surveyors prepare, virtually all contain the boundary lines of the parcel in question, and the boundary lines relate to all the other data on the map. The boundary defines the relationship of the subject parcel to the adjoining parcels. The boundary purports to represent the limits of the vesting deed of the subject parcel. Even with topographic surveys (which contain a wealth of information about

1. An earlier version of this article appeared in 18(5) PROBATE & PROPERTY (Sept./Oct. 2004).

utilities, terrain features, structures, and easements), none of the data shown on the map makes any sense without that crucial spatial relationship to a boundary.

And yet most state laws do not even begin to spell out the ingredients of a properly resolved boundary. California law, for example, sums up the entire matter in nineteen words. In discussing the technical requirements of a survey map, the statute states that the survey shall show "[t]he relationship to those portions of adjacent tracts, streets, or senior conveyances which have common lines with the survey." (CAL. BUS. & PROF. CODE § 8764).

It is, of course, impossible to condense the entire accumulated wisdom of boundary law and precedent in a single sentence. All the good intentions of those who wrote the law aside, it is simple human nature that if the law does not lay out the principles to be followed, then a great deal can be saved by choosing the path of least resistance. Many surveyors, even county and city surveyors, have reported that the only requirement for "treating adjoining deeds" is that the book and page reference be noted on the map. But to conduct a thorough study of the deeds for the subject parcel and each adjoining parcel to make sure that there are no conflicts between any of them takes many hours, sometimes days, and clearly would result in a much larger bill for the client.

## I. Definition of a Boundary Resolution

The resolution of a boundary refers to the entire process of analyzing all available, pertinent information relating to the location of the boundary of a given subject parcel *in its relationship to the boundaries of the parcels (and/or rights-of-way) that surround it*, with an eye to determining whether there exist any inconsistencies or incompatibilities between any of those boundaries.

A parcel is rarely, if ever, an island unto itself. Its existence and location are integrally related to the parcels that surround it. "Resolving" a boundary necessitates a consideration of the parcels and rights-of-way that border it on every side. A good map will present the evidence and analysis relied on to conclude where these boundaries lie.

## II. What to Look For

Some surveyors might be lazy, or cheap, or ignorant, or perhaps even fraudulent. Given the varying competence of surveyors, and given that some state laws do little to enforce a standard and competence, what can individuals in the allied professions do when examining a map to get a better idea of how good its boundary is? What follows are suggested guidelines. Remember that surveying is, indeed, a highly complicated art and science, and that this short article cannot do justice to its many subtleties. Nevertheless, some general principles should be kept in mind when examining a map.

Most non-surveyors do not know what questions to ask when they look at a map. If the surveyor who prepared the map is available, ask him or her to describe the evidence used to place the boundary of the subject parcel where it is located on the map.

Every angle point, or every beginning and ending point of a particular line, should be justified on the map. The map reader should realize that the surveyor should have a reason for the location of every angle point and every end point of a line or curve. Some surveyors are more conscientious than others as to how clearly they display this data, but the data should be there in some form. If the map cannot be adequately interpreted by itself, then ask the surveyor to explain the rationale behind his or her resolution.

Usually, the best justification for locating a line will be that the surveyor found a good monument in the field that marks one or both ends of the line (or curve) in question. If a map symbol indicates a "found monument" at each end of a line, then that means the surveyor has decided those monuments are good and will control the position and length of the line. These monuments can serve this purpose because they have a good pedigree (for example, they have been accepted by many surveyors on prior recorded maps), or because the surveyor found a deed that specifically mentions the existence of the monuments. Other justifications can be used for accepting a monument, but the point is that it was found in the field and, in the surveyor's professional judgment, it should be used to control the position of the line.

The more found monuments there are at the angle points of the parcel boundary, the more solid is the resolution of the map. But that is not always the case. What controls a boundary if there are no found monuments anywhere around the boundary?

A line can be located by other good reasons. It might be because a remote monument was found down the street and the surveyor held a record distance from that monument to locate a point in the subject parcel boundary. This is an acceptable alternative justification for setting the position of a line. It has two elements to it: (i) a monument was found, and (ii) a record distance was used to measure from that monument to the point that the surveyor needed to locate. These are good reasons for making a decision about a particular line or angle point.

But what about that "record distance" mentioned above? How would the map reader know that the surveyor had used a record distance to locate a point on the subject parcel boundary? A typical way would be to show that record distance inside parentheses, with a number or letter behind it, like this: (135.34')R3. The "R3" refers to a record reference. A table of Record References will appear somewhere on the map, and the surveyor is calling attention to the third reference in the list. The table entry for R3 might be the book and page of a recorded deed. It might be a previously recorded map of some type. Thus, what the surveyor is saying is, “I found the distance of 135.34 feet mentioned in a deed and determined that this distance should be held in order to arrive at the position of one of the angle points in the boundary of the subject parcel.”

Such language is drafting shorthand used by surveyors. Drafting styles vary around the country, and there are many other ways of indicating that the surveyor has “used” (relied upon) a record distance to control a point or line.

Another possibility is to add an explanatory note. Suppose that some monuments were found and used to establish the right-of-way line along a street. Then, coming off of this line, the map shows one of the side lines of the subject parcel. Questions arise: What angle is formed between this side line and the right-of-way line? Why does the side line have the bearing (the angle) that it does? Sometimes surveyors will draw an arc between the two lines and add a note stating: “Record angle held, per R2.” This indicates that, since a monument was not found at the back end of the side line, the surveyor determined that the next best justification for establishing the bearing of the side line was to “hold” a record angle between the two lines according to some other recorded document or map.

Sometimes a note will be so long and complex as to deserve a special statement off to the side of the map. A surveyor may have gone through a lengthy analysis of a line or point and needed to summarize his or her thinking by describing in some detail how that part of the boundary was resolved. Such a summary is always a good sign, because it shows that the surveyor takes the resolution of the boundary seriously and wants to make sure that anyone examining the map can understand that resolution.

No matter which method is used, one should be able to decipher the rationale behind each and every decision made in the location of a line, curve, or point. The more explanatory data on the map, the better the map and the more defensible its resolution.

If a map reader is examining a map and cannot tell why a line was located where it was, there is nothing wrong in calling the surveyor to ask for an explanation. It may be uncommon, but it should be no less acceptable than if a patient were to ask his or her doctor to explain exactly why a certain prescription was given or why a test was being ordered. If the lawyer is using a map that has important implications for a project or a case, he can, and should, ask to meet with the surveyor to secure a complete breakdown of the reasoning that went into the boundary shown on the map.

Look at it this way: a boundary has to be built upon a strong foundation, and ultimately that foundation is based upon the monuments that were located in the field. The perfect situation would be for monument to be found at each angle point going around the boundary. But if only some (or even none) were found along the boundary, then remote monuments come into play and the boundary can be built from them. As the web of parcel lines is constructed from the remote monuments, the surveyor calculates the way back to the subject parcel and eventually is able to position the boundary from them. But each step along the way needs to be justified, and if that justification is not obvious on the map, asking the surveyor to explain things makes sense.

Alternatively, seek a second opinion. As with a doctor's diagnosis, a map depicts the professional opinion of a particular surveyor on where a parcel lies relative to the parcels surrounding it and relative to the monuments that were located in the field. Many times this author has been brought a map and asked to analyze and explain what the surveyor did. It is not that hard for an experienced surveyor to study the information on a map and then to describe what it says about how the other surveyor resolved the boundaries it shows. Summarizing the strengths and weaknesses of the map is also possible.

The best maps are those that faithfully present a summary of the evidence found and all the key decisions that link the subject parcel boundary to that evidence. Because the record of these decisions is often lacking on maps with which they work, lawyers are forced to question the quality of their resolutions. Is it because the surveyor, for whatever reason, elected to simplify and therefore did not add enough notes to explain the resolution to the reader? Or is it because the surveyor is doing shoddy work and is hiding behind the brevity of the law, thus producing a map with little or no justification for the decisions it represents?

## III. Conclusion

Ideally, a map should show the basis for the surveyor's determination of the boundaries. The basis, however, is not always evident from looking at the map. Because the law does not require the surveyor to show a complete resolution on his or her map, how is the map reader to know that the map is accurate? The answer is that one cannot know the answer. And this realization is what prompts the recommendation that the map reader go beyond merely trusting that a surveyor does competent work. If a map appears to be lacking in documentation or analysis, the lawyer should contact the surveyor who prepared it and ask him to fill in the blanks and, depending on the circumstances, consider obtaining a second opinion from another surveyor.

# Advanced Survey Examination Issues[1]

3

Richard F. Bales

## I. Introduction

To a large degree, the process of examining surveys of commercial property is very similar to the examination of residential plats of survey. Is there a building line violation? If so, can it be endorsed over on the title insurance policy? If there is an encroachment onto adjoining property, can it be endorsed over? Although an encroachment onto the land in question cannot be endorsed over on an owner's title insurance policy, will the parties sign an easement or license agreement? All of these items must be considered when reviewing the commercial survey.

But the review of surveys of commercial property is more than merely looking for encroachments and building line violations. The attorney must also be aware of a myriad of other matters, items, and concepts, some of which are rather subtle. These advanced survey examination issues will be discussed in this article.

1. An earlier version of this article appeared in PROBATE & PROPERTY (Sept./Oct. 2004) and was adapted from RESIDENTIAL AND COMMERCIAL REAL ESTATE SURVEYS IN ILLINOIS by Richard Bales and Michael Filipski, published by the Illinois Institute for Continuing Legal Education, 2004.

## II. The Survey Seal, Certification, and Notes

When reviewing a plat of survey, the attorney should not ignore the surveyor's seal, certification, and any notes that may be appended to the plat. The attorney should be concerned if a plat of survey does not have a raised or inked seal or if the seal looks "fake." On occasion, surveyors' clients have altered a completed survey or plat of subdivision, photocopied the plat, and then passed it off as the original. To combat this problem, many surveys will contain a disclaimer, indicating that they are not valid unless they have a raised seal. Some states may have legislation that addresses the problem of plat alteration. For example, Illinois law provides that only the surveyor or his or her designee can record a plat of subdivision.

A suggested survey certification is included in the 2011 ALTA/ACSM land title survey standards. The certification reads as follows:

"This is to certify that this map or plat and the survey on which it is based were made in accordance with the 2011 'Minimum Standard Detail Requirements for ALTA/ACSM Land Title Surveys,' jointly established and adopted by ALTA and NSPS, and includes Items _____ of Table A thereof. The fieldwork was completed on _________." (Note that appended to the 2011 ALTA/ACSM land title survey standards is a "Table A" of "Optional Survey Responsibilities and Specifications." This is a list of twenty-one optional items which, if requested, can be included on the land title survey. The list includes such matters as the gross land area of the property, the current zoning classification of the property, and the exterior dimension of all buildings located on the property.)

The attorney should always read the survey certification of any survey he or she is reviewing and should examine closely any certification that deviates substantially from the ALTA/ACSM certification. Generally speaking, the plat of survey should contain words to the effect that the surveyor has *surveyed* Blackacre. If instead the plat indicates that the surveyor has *platted* or *located* Blackacre, the attorney should ask the surveyor to clarify the certification.

Consider, for example, the following survey certifications: "I, ______, hereby declare that this plat is a correct representation of a survey made from existing plats and records." This certification suggests that the surveyor was never in the field at all. It is clearly ambiguous and should be explained. Another variation of this certification is the following: "I, ______, hereby state that this map or plat was based upon existing maps, plats, and surveys previously prepared by this office to ALTA/ACSM standards."

A nonstandard survey certification is not necessarily an indication that the surveyor is trying to limit his or her liability. It is possible that because of either time or cost considerations, the surveyor's client has asked the surveyor to perform nonstandard services.

For example, a surveyor wrote the second of the above two certifications after the surveyor was given only two days to survey a large tract of land. The surveyor had surveyed the property a year earlier but was not able to resurvey the property in the time allotted. As a result, the surveyor updated the survey without performing additional fieldwork. Unfortunately, a title company might not accept a survey with such a certification unless the company is provided additional assurances.

Finally, consider this certification: "I, _____, a professional land surveyor, hereby certify that the parcel plat hereon drawn has been prepared under my direction and this plat is a correct representation of the property described therein."

When the surveyor was asked to explain this certification, he replied that his "survey" was actually a drawing based on a proposed plat of subdivision that he had prepared for a developer. Again, a title company will probably not accept such a certification.

Any "surveyor's notes" should be read carefully. Most of the time, these notes will aid the attorney in better understanding any issues or problems that the surveyor encountered while surveying the property. In addition, surveyors will often include information for the convenience of the client. This additional information might include the common address of the property or the identification of title commitment Schedule B exceptions. If the survey legal description differs from the legal description in the title commitment, the surveyor might add a statement explaining this discrepancy.

But it is possible that a surveyor will include a note that limits the assurances of the survey certification. Again, this may be due to the demands of the surveyor's client. For example: "Observable evidence of utility easements is not shown pursuant to the client's request." Clearly, this note almost emasculates the certification of the land title survey.

On the other hand, surveyors will sometimes add notes to their surveys that limit their liability under the land title survey certification. Consider the following survey note:

"No statement is made concerning subsurface conditions, the existence of underground or overhead containers, or facilities that may affect the use or development of this tract. The underground, surface, and overhead utilities have been located from field survey information and existing drawings. The surveyor makes no guarantees that the underground, surface, and overhead utilities shown comprise all such utilities in the area, either in service or abandoned. The surveyor further does not warrant that the underground utilities shown are in the exact location indicated, although he does state that they are located as accurately as possible from information available. The surveyor has not physically located the underground utilities. The surveyor makes no claim to specific knowledge regarding the usage or type of overhead wiring shown hereon."

It appears that by adding this note to his survey, the surveyor has placed himself in an interesting dilemma. Generally speaking, the land title survey standards require that the surveyor show only observable evidence of utilities on the plat of survey. But once the surveyor decides to show the location of underground utilities, what is the liability of the surveyor?

This surveyor has attempted to limit his liability with this note. But the note also refers to aboveground utilities, and by adding this reference, the note takes away some of the assurances of the land title survey certificate. For example, by indicating that "no statement is made concerning subsurface conditions, the existence of underground or overhead containers, or facilities that may affect the use or development of this tract," it appears that there may be other overhead utilities on the property that are not shown on the survey. Such a statement dilutes the survey certification because paragraph 5(E)(ii) of the survey standards requires the surveyor to show "Evidence of easements or servitudes . . . observed in the process of conducting the survey."

Instead of drafting a note that may cause problems at closing, this surveyor should instead consider including a "Table A" reference to item 11 in the survey certification. Item 11 requires the surveyor to show on the survey the following: "location of utilities . . . existing on or serving the surveyed property as determined by: observed evidence together with evidence from plans from utility companies or provided by client, and markings by utility companies, and other appropriate sources (with reference to the source of information)." Item 11 sufficiently protects the surveyor, as it limits the surveyor's liability regarding the location of underground utilities to observable evidence of the utilities and to plans provided by the client and other sources.

## III. Fences

Boundary lines are nonphysical lines that have no width. On the other hand, fences are physical objects that do have a width. Many surveys will indicate that a boundary fence is "on line," that is, the fence has been constructed along the boundary line of the property. But how can a fence be "on line" without some portion of the fence being across the boundary line and encroaching onto adjoining property?

A fence is "on line" when any visible part of the fence is on the boundary line of the property. If, for example, three-fourths of the width of a board fence is on adjoining property but the remaining one-fourth is on the land being surveyed, that fence is still "on line." Only when the fence is completely on the adjacent property, with no part of the fence "on line," is there a fence encroachment. (See Stephen V. Estopinal, A Guide to Understanding Land Surveys 132–34 (1989).

Note that when surveyors survey land, they will usually not know who owns a fence. Therefore, fences will not be labeled as "encroachments" on the plat of survey. That is, there will not be a statement on the survey that "A's" fence encroaches onto "B's" property. Rather, surveyors will usually simply depict the location of the fence and indicate the distance from the fence to the boundary line of the land.

Residential legal descriptions may simply consist of a lot number and a subdivision name. But legal descriptions for commercial property can often be much more complex. A metes-and-bounds description of unsubdivided property might be several pages long, especially if it is a large tract that falls in two or more sections of land.

When reviewing a survey of unsubdivided land (e.g., a quarter-quarter section, or approximately forty acres), the attorney may want to question a survey that discloses a boundary fence that is "on line" relative to an interior section line, such as a quarter section line or a quarter-quarter section line. (See figure 1.) These interior lines were never surveyed by the original government surveyors; they can be located only by protracting (i.e., by measuring and then establishing on the ground) the original section. Many times, landowners would construct boundary fences by running a line inward from the north, south, east, or west quarter section corner. (See figure 1). But because these interior lines were never surveyed, these interior boundary fences were not always constructed on these interior lines. A survey that shows a fence "on line" relative to an interior section line might be an indication that the surveyor merely assumed that the fence marked the location of the interior line.

On the other hand, the exterior section lines and all eight exterior section corners (see figure 1) *were* plotted by the government surveyors. Many times, fences located along these exterior lines are the only remaining evidence of these section lines. It is possible that a fence constructed along an exterior section line accurately marks the location of the section line. Nonetheless, the attorney may want to question a survey that shows a fence "on line" relative to an exterior section line or discloses a fence along an exterior section line but gives no dimensions relative to the location of the fence and the section line. Again, it is possible that the surveyor assumed that the fence marked the exterior section line without performing the necessary fieldwork to verify the location of the section line.

Although fences constructed along interior section lines may not be accurate evidence of the location of these section lines, the attorney and the surveyor should not automatically dismiss these fences as irrelevant and unimportant. As shown in figure 1, these fences may tie into the north, south, east, or west quarter section corners, which were established by the original government surveyors. Although these fences may not be constructed along the interior section lines, they may aid in the location of these exterior section corners.

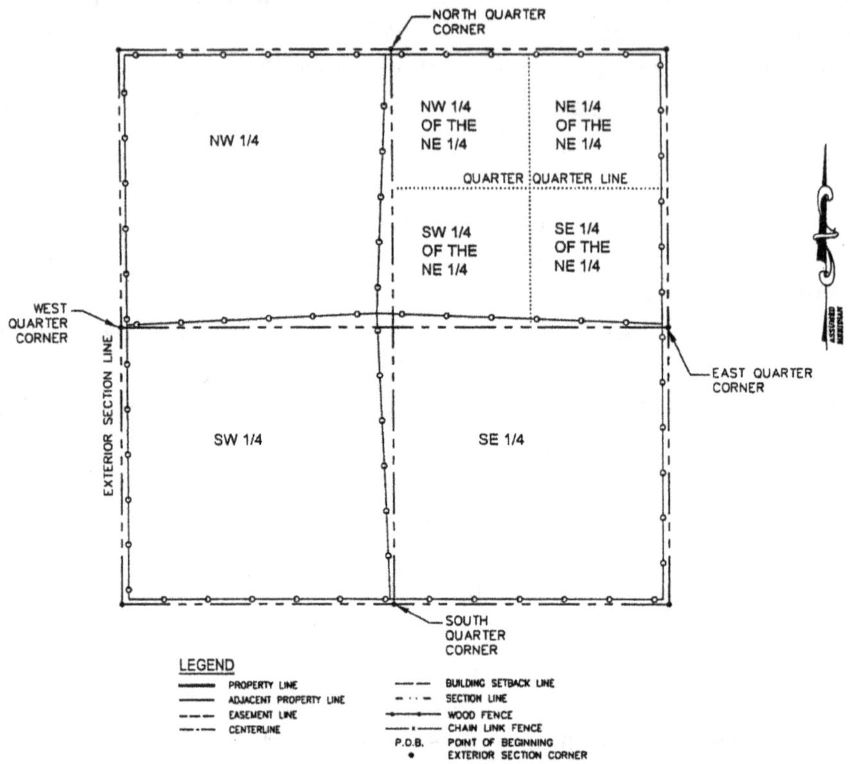


Interior section lines such as quarter section lines and quarter quarter lines were never surveyed by the original government surveyors. These lines can be determined only by first surveying the original section of land and then establishing these interior lines.

On the other hand, the exterior section lines *were* surveyed and set by the original government surveyors. Many times fences located along these section lines are the only remaining evidence of these lines.

Here, a fence has been constructed along the exterior lines of the section. Although a second fence correctly ties into the north, south, east, and west quarter corners of the section, this fence does not fall along the interior section lines.

**Figure 1.**

## IV. "Actual" and "Occupied" Boundary Lines

The attorney should question a survey that discloses two identically named section lines or record monuments. For example: the south line of the northwest quarter and the south line of the northwest quarter *as occupied* (see figure 2). Strictly speaking, this notation is incorrect; there can be only one section line, as this is a *record monument* or *title line* that never changes location. On the other hand, the location of *ownership lines* can change because of such legal doctrines as adverse possession.

Example: Jane owns the northwest quarter of section 3. The four boundaries of this quarter section are title lines; the location of these lines cannot change.

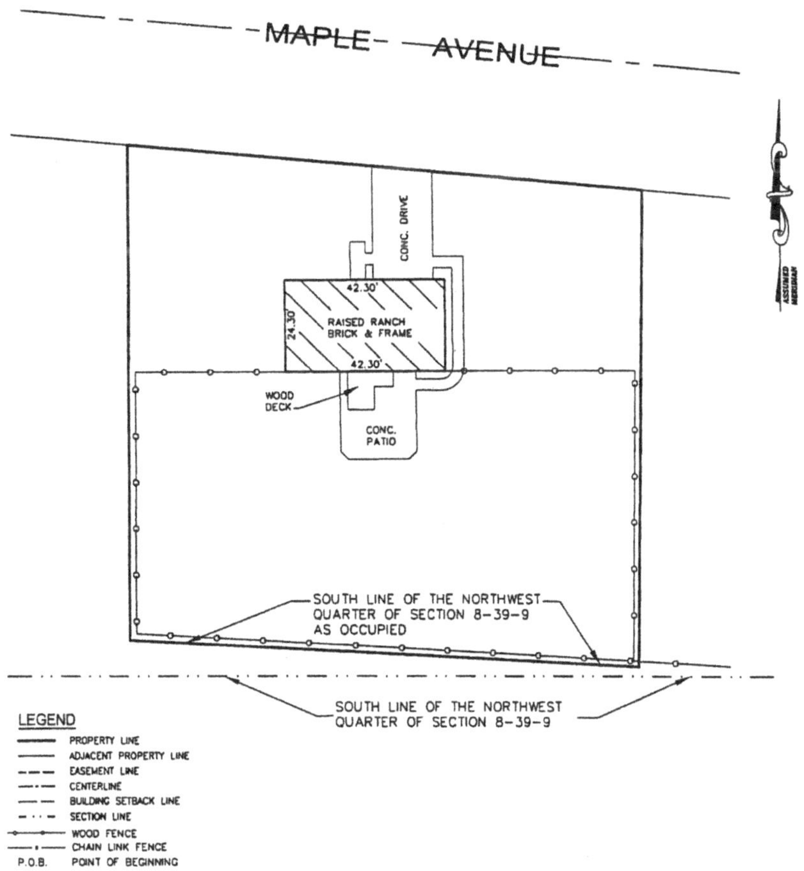


Note the two south lines of the Northwest quarter of Section 8, Township 39 North, Range 9; one of them is the south line "as occupied."

**Figure 2.**

Jane fences in and adversely occupies a portion of the adjoining quarter section, the northeast quarter of section 3, for twenty years. It is possible that because she has adversely occupied a portion of the northeast quarter for the statutory period, her ownership lines have changed to include this additional land.

Although the identification of section lines and lot lines "as occupied" is, strictly speaking, incorrect, a survey disclosing the use of this term is probably not defective. For example, it is possible that the surveyor who surveyed the property shown in figure 2 realized that there could be only one south line of a quarter section, but wanted to point out that the field location of the fence traditionally thought to be located along the quarter section line and the location of the actual quarter section line, as determined by the surveyor, are not in the same place. When used in this manner, the identification of a section line or lot line is an acceptable "phrase of convenience." Nonetheless, the attorney should not assume this to be the case, but instead should ask the surveyor to explain his or her use of the term.

Consider the survey of the following described property (see figure 3):

"Beginning at the southwest corner of Blackacre Subdivision; thence south 5 degrees east to the south line of the southwest quarter of Section 1 *as occupied*; thence east along said south line to the west line of Plank Road; thence north 20 degrees east along said west line to the southeast corner of lot 10 in Blackacre Subdivision; thence west along the south line of said Blackacre Subdivision to the point of beginning."

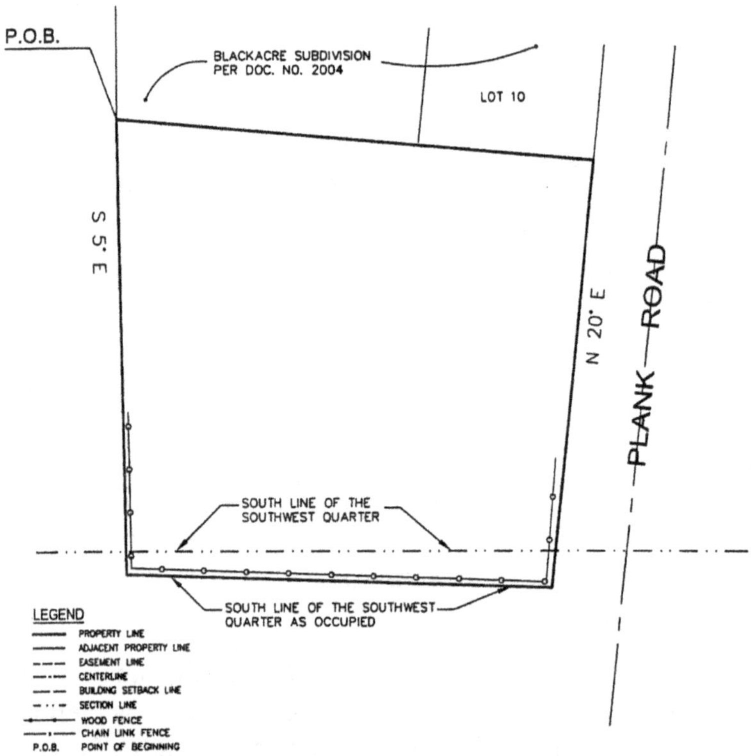


The legal description of this tract of land is as follows:

"Beginning at the southwest corner of Blackacre Subdivision; thence south 5 degrees east to the south line of the southwest quarter of Section 1 *as occupied*; thence east along said south line to the west line of Plank Road; thence north 20 degrees east along said west line to the southeast corner of lot 10 in Blackacre Subdivision; thence west along the south line of said Blackacre Subdivision to the point of beginning."

Does the owner of the property actually own the land between the two "south" lines, or is the owner merely adversely occupying this property? This is an issue that the title insurance company and the attorney for the owner must address.

**Figure 3.**

This is another example of a survey that discloses two south lines—one actual south line and a south line "as occupied" by a fence. But here, the surveyor has actually amended the legal description to reference the south line of the southwest quarter, "as occupied." Again, perhaps the surveyor knows that there can be only one south line of a quarter section but wanted to point out that the fence, which third parties thought was built along the quarter section line, and the actual quarter section line are not in the same location. In this instance and in the previous example, the attorney will have to work with the surveyor and the title company to resolve this apparent conflict.

How can this matter be rectified? The surveyor should examine the legal descriptions of deeds of the adjoining property and compare the descriptions to the legal description of the land in question. Are there any gaps or overlaps? The surveyor should talk to other surveyors who have worked in this area and examine their surveys, if possible. How did these surveyors address this problem? Did their legal descriptions use this occupation line as a monument? Have improvements been constructed along this occupation line? Because there can be only one quarter section line, the surveyor may want to rewrite the legal description. For example, the surveyor could either eliminate the reference to the south line of the southwest quarter, as occupied, or reference the location of the "occupied" south line relative to the location of the "real" south line.

## V. "Record" versus "Measured" Distances of Land Boundaries

Plats of surveys will sometimes show "record" and "measured" distances. That is, a lot line might have a platted or "record" dimension of a certain number of feet. When the surveyor goes out into the field and actually measures this lot line, the surveyor discovers a "measured" distance that may be more or less than the actual record distance. This difference between record and measured distances is sometimes due to the greater accuracy of new technology.

But the attorney should question large variations of "record" and "measured" distances. For example: A surveyor is hired to survey a lot in a small five-lot commercial subdivision. The record dimension of each lot is 50 feet in width, for a total of 250 feet. The surveyor goes to the property and locates iron pipes set near each corner of the lot in question. The surveyor then measures the distance between the two pipes and determines the width of the lot to be 49.5 feet. The surveyor indicates on the plat of survey that the width of the lot is "50 feet record, 49.5 feet measured." But it is possible that if the surveyor had measured the entire width of the subdivision, the surveyor would have found original lot corners indicating that the subdivision was indeed 250 feet in length and that the lots were all 50 feet in width. Such a finding would suggest that at least one of the iron pipes the surveyor measured from was in the wrong location.

Experienced surveyors know that a pipe in the ground is not sacrosanct. Lot corners are often moved; common culprits are utility companies, cable companies, and fence installers. Also, subsequent surveyors (usually called retracement surveyors) may not find an original lot corner monument in the ground when performing their fieldwork. Rather than perform the necessary fieldwork to properly restore the lost corner, they may instead place a new pipe in the ground at what they feel is the lot corner. Indeed, surveyors often tell of instances where two or more pipes can be found at a lot corner. A measurement taken from the wrong pipe can obviously produce a difference between record and measured distances.

So how much of a difference between record and actual measurements is questionable? Even three-tenths of a foot in a 100-foot length might be considered a substantial deviation in some cases. Although a marked difference between record and measured distances does not automatically mean that the survey is defective, the attorney may want to ask the surveyor to explain, if possible, the reason for the difference between the two distances.

But the attorney should realize that the issue of record and measured distances is extremely subjective. For example, sometimes the age of the subdivision is a factor. That is, there might be a rather large (but still legitimate) difference between record and measured distances in a very old subdivision. Unfortunately, there is no "table of tolerable deviation" to which the attorney can refer. Every situation is different.

When land surveying was in its infancy, the Gunter's chain and compass were used to measure distances and bearings. Surveyors today have electronic wonders like GPS and EDMs at their disposal. Is it any wonder that there are differences between record and measured distances and bearings? But should these differences be memorialized in the public record? That is, should legal descriptions be amended to reflect both the record and measured distances and bearings?

For example, consider this record description: "thence North 30 degrees 46 minutes 18 seconds East 246.00 feet to the east line of the northwest quarter of Section 6; thence . . . " After the surveyor surveys the land and discovers a difference between the record distance and the measured distance, the surveyor revises the description to disclose both these record and measured distances: "thence North 30 degrees 46 minutes 18 seconds East 246.00 feet record (246.59 feet measured) to the east line of the northwest quarter of Section 6; thence . . . "

This issue of amending legal descriptions to reflect differences between record and measured distances is an extremely controversial one, especially among land surveyors. Some people feel that the record description is virtually sacred and that any revised description should contain the original record distances and bearings. On the other hand, many people feel that if the legal description contains references to artificial, natural, or legal monuments, the legal description can be amended without including the old record distance or bearing, as these monuments take precedence over the original distance or bearing. Note, for example, how the second description has been revised to delete the references to the original distance: "thence North 30 degrees 46 minutes 18 seconds East 246.59 feet to the east line of the northwest quarter of Section 6; thence . . . "

## VI. The Apportionment Rule, Also Known as the Proportional Measurement Rule

One possible explanation for differences between record and measured lot distances may be due to record and measured differences between subdivision corners or block corners. When a surveyor uncovers original subdivision monuments or is able to reconstruct original monumentation and thereby discovers legitimate differences between record and measured distances that are between subdivision corners or between block corners, the surveyor will use the Apportionment Rule to apportion these differences among the various lots. That is, the excess or deficiency that the surveyor discovers is added or subtracted to each lot in direct proportion to the size of each lot. Thus, if lot 1 is 100 feet wide and lot 2 is 200 feet wide, lot 1 gets one-third of any excess or deficiency and lot 2 gets two-thirds of any excess or deficiency. This only makes sense; since lot 2 is twice as big as lot 1, lot 2 gets twice as much excess or deficiency as lot 1. Rarely, however, are excess and deficiency problems this easy to solve.

The land surveyor will use the Apportionment Rule to *identify* lot boundaries and not *change* lot boundaries; therefore, the surveyor can unilaterally apply the rule without first requiring deeds from the lot owners.

Example: Blackacre is a four-lot platted industrial park. The subdivision is rectangular in shape. The lots have record widths of 50 feet, 100 feet, 100 feet, and 350 feet. Thus, there is a record distance of 600 feet. However, the surveyor locates the four original subdivision corners and determines that the measured length of the subdivision is actually 603 feet. There is 3 feet of excess within the subdivision. This 3 feet of excess length must now be apportioned between all four lots. The procedure is as follows:

First, take the excess (or deficiency, as the case may be), which in this example is 3 feet, and divide it by the record subdivision distance (or block distance, if one is apportioning the distance within a block):

$3 \div 600 = 0.005$; this figure is called the *constant.*

Now multiply the constant (0.005) by the record width of each lot. This is the proration that is added to each lot. The record distance plus this proration is the measured distance:

$0.005 \times 50 = 0.25$; therefore, the 50-foot lot has a measured distance of 50.25 feet.

$0.005 \times 100 = 0.5$; therefore, the 100-foot lot has a measured distance of 100.5 feet.

$0.005 \times 100 = 0.5$; therefore, the 100-foot lot has a measured distance of 100.5 feet.

$0.005 \times 350 = 1.75$; therefore, the 350-foot lot has a measured distance of 351.75 feet.

Note that if these measured distances are added together, the total is 603 feet, which is the measured distance of the length of the subdivision: $50.25 + 100.5 + 100.5 + 351.75 = 603$.

However, if there is a street or alley within the block or subdivision, one does not prorate the width of the right-of-way. The surveyor's rule (which dates back to old English common law) is: streets and alleys get their full measure; the "sovereign" cannot be dispossessed of its land.

Example: A block in Whiteacre Subdivision has four lots; each lot is 50 feet wide. A 16-foot alley runs through the middle of the block. The surveyor finds the original block corners and determines that although the record distance of the block is 216 feet, the measured distance is actually 214 feet. That is, there is a 2-foot deficiency within the block.

But because "streets and alleys get their full measure," the deficiency is apportioned only to the four lots and not to the 16-foot alley. The width of the alley remains the same. Therefore, in order to compute the constant, the surveyor must first deduct the width of the alley before dividing the deficiency by the (adjusted) record block distance: 216 (record block distance) less 16 (width of alley) is 200 (adjusted block distance).

Again, divide the deficiency by the (adjusted) record block distance: $2 \div 200 = 0.01$.

Now multiply the constant by the record width of each lot: $0.01 \times 50 = 0.5$.

Now deduct 0.5 from the width of each lot: $50 - 0.5 = 49.5$ feet.

The measured distance of the block is 214 feet. Each lot in the subdivision has a measured distance of 49.5 feet. The width of the 16-foot alley is not adjusted; it remains the same. Therefore: $49.5 + 49.5 + 49.5 + 49.5 + 16 = 214$.

Note that the above examples are given only to illustrate the concepts set forth in this section. The differences between "record" and "measured" distances within subdivisions are not usually as severe as the differences in these examples.

## VII. Measured Distances from Improvements to Property Lines

When reviewing a survey, the attorney should be concerned if there are no "ties," or measured distances, from any building on the property to the boundary line of the land. This could be an indication that the surveyor did not perform all the necessary fieldwork. Similarly, the attorney should question a plat of survey that depicts a fence along a property line but does not disclose either that the fence is "on line" or the distance from the fence to the property line.

Note that paragraph 5(D) of the 2011 ALTA/ACSM land title survey standards requires that the surveyor show "the location of all buildings on the surveyed property shown perpendicular to the nearest perimeter boundary lines. . . " If the surveyor has not shown this information, it is possible that the surveyor's client requested that this information not be disclosed, perhaps due to either time or cost considerations. If this is the case, however, the surveyor will probably qualify the survey certification accordingly. The title company may question any amended survey certification.

## VIII. Evidence of Adverse Possession

Figure 4 indicates that the owner of the adjoining property to the north has been mowing and landscaping a portion of the land in question. This owner has also placed plastic playground equipment on the property. Along the east line, an owner to the east has placed planting material in the southeast corner of the land. Another owner has dumped yard waste on the land.

Could this be evidence of adverse possession? In *Joiner v. Janssen*, 421 N.E.2d 170 (Ill. 1981), the Illinois Supreme Court found there were sufficient acts of adverse possession in that the "[p]laintiffs mowed the grass on the 14-foot strip in question, raked leaves, planted and removed trees, bushes and flowers, gave away trees, bushes and flowers from the land as gifts, buried their pet dog on the strip when it died, shoveled snow from the walk in front of the strip, and generally maintained the property . . . ." *Id.* at 174.

Note that the surveyor has a duty to show evidence of adverse possession on the land title survey. Paragraph 5C(i) of the 2011 ALTA/ACSM land title survey standards provides that the surveyor shall disclose "the character and location of evidence of possession or occupation along the perimeter of the surveyed property . . . "

So how can a title exception relative to the possible possessory rights of adjoining neighbors be waived from the title commitment? In the above instance, the attorney for the seller went to the site and talked to the owners of the land to the north and to the east. Neither neighbor was attempting to acquire any ownership rights to the land. As indicated by photographs taken by the attorney, the property was simply a huge tract of weeds and brush that the two men were using as a dumping ground for yard waste. These pictures, together with the attorney's letter of explanation, convinced the title company that it need not be concerned about these neighbors attempting to acquire an ownership interest in the land.

## IX. Unrecorded Easements

As noted above, the examination of surveys of commercial property is more than just looking for encroachments and building line violations. Even such seemingly

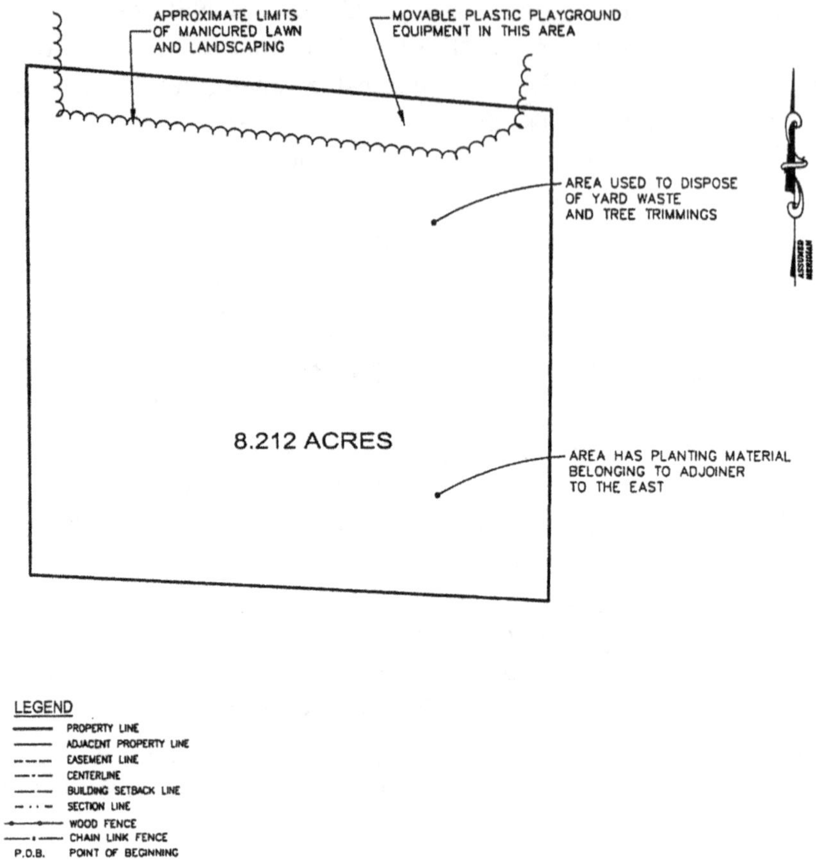


The surveyor completed a land title survey of this vacant tract of land. Pursuant to paragraph 5(f) of the ALTA/ACSM land title survey standards, the surveyor showed this evidence of adverse use of the property.

**Figure 4.**

simple matters as unrecorded easements take on a whole new dimension when working in the commercial arena.

The survey of an industrial building at figure 5 indicates that an adjoining commercial structure was built along the east line of the property in question. Note that the surveyor has indicated that two doors along this east line swing out and thus over onto the surveyed property. As the easterly portion of this property is a parking lot, it appears that the owner of this building (and possibly the owner's customers) has been walking back and forth over the property in question in order to enter and exit this east building. If this is the case, then it is possible that the land in question is subject to at least two unrecorded easements—an easement for the two doors that swing out and over onto the land and an easement

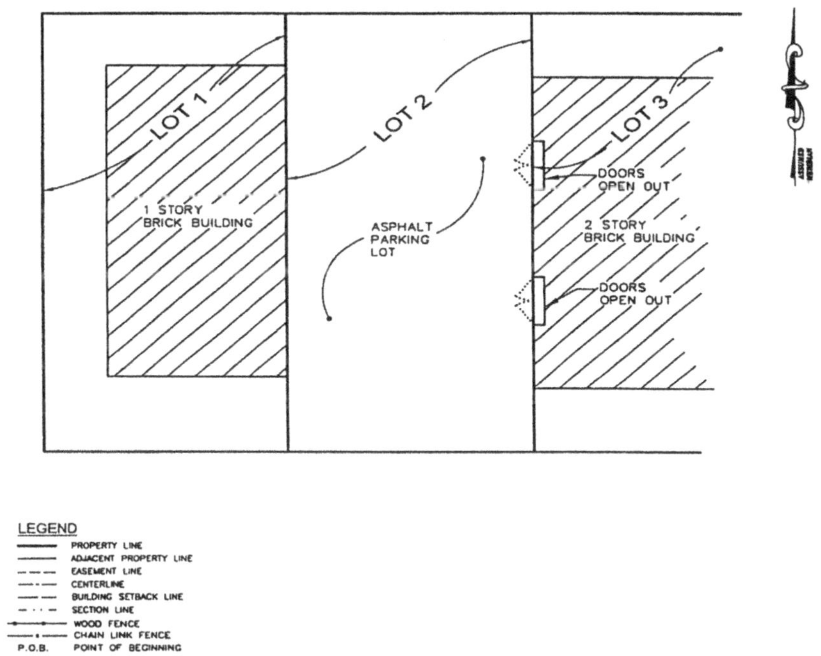


The property in question consists of two lots, lots 1 and 2. A building is built on lot 1; lot 2 is an asphalt parking lot. Another building has been constructed on adjoining lot 3. This building has been built up to the lot line between lots 2 and 3. The two doors of the building swing out and onto the parking lot.

**Figure 5.**

for pedestrian access. And are the owner and the owner's customers parking their cars on this parking lot? If so, there may be a third easement.

Is there any way that a title insurance company might underwrite these survey matters to the satisfaction of its insured's counsel? With appropriate affidavits wherein, for example, the owner of the adjoining property affirms that he and his customers do not normally use the doors, the title company might be willing to raise an exception, but also include an endorsement indicating that, as of the date of policy, there has been no easement created as a result of the doors swinging out onto the insured land. Such an endorsement might read as follows:

> The Company hereby insures the Insured against loss or damage that the Insured shall sustain by reason of the entry of any final judgment by a court of competent jurisdiction, determining that as of the date of policy the owner of the adjoining land to the east has acquired an easement (by prescription or otherwise) over the land described in Schedule A, as noted

in Schedule B at exception _____ and as shown on the land title survey of _____, dated _____, job number _____.

The title company will probably want the seller's assurances that he or she will pay all defense costs in the event that the neighbor later litigates the right to use the doors for ingress and egress. These assurances are usually memorialized in a document called either a personal undertaking or an unsecured title indemnity. In the event that a court later determines that the neighbor had already established an easement over the insured property as of the date of the policy, the title company would have liability under the endorsement.

Note that the endorsement would be applicable only to the possible creation of an easement prior to the date of the title policy. The endorsement would not protect against the court-ordered determination that an easement was created sometime after the date of the policy.

The 2011 ALTA/ACSM land title survey standards impose a duty on the surveyor to disclose these doors, even though they are appurtenant to a building that is located on adjoining property. Paragraph 5(C)(iii) of the standards provides as follows: "Without expressing a legal opinion as to the ownership or nature of the potential encroachment, the evidence, location and extent of potentially encroaching structural appurtenances and projections observed in the process of conducting the survey, such as . . . windows and doors that open out . . . [shall be shown on the plat of survey]."

Note that this paragraph also offers an enlightening observation on what the surveyor should *not* show on his or her survey. This paragraph provides that the surveyor should disclose encroachments "without expressing a legal opinion." Thus, the surveyor will indicate on the survey of the land that the doors "swing out" onto the adjoining property without adding that "the neighbor to the east has acquired an easement by prescription over a portion of the surveyed property."

Surveyors will normally avoid making such legal conclusions. Most surveyors are not lawyers and therefore do not have the legal training to make such statements. In addition, it is possible that the provisions of their errors and omissions insurance mandate that they show only *facts* on their surveys and that they do not draw legal *conclusions* from these facts. Therefore, surveyors may show physical evidence of encroachments or easements, but they will not make any conclusions relative to this evidence. For example, although the surveyor may show an encroaching building onto adjoining property, the surveyor will not state on the survey that "the owner of the land described on this plat now owns by adverse possession the land underneath the encroaching building to the east."

As mentioned above, although a surveyor may show an encroaching fence on the plat of survey, the surveyor will not normally indicate who (e.g., the landowner to the north) owns the fence. Although the surveyor may show a vacated right-of-way appurtenant to other surveyed property, the surveyor will not state that "upon vacation of the right-of-way, the owner of the adjoining property acquired a fee simple interest in said right-of-way." Although a surveyor may indicate on the plat of survey that a driveway encroaches onto the land in question, the surveyor will not label the encroachment a "prescriptive easement."

## X. Conclusion

This article only taps the surface of advanced survey issues that the attorney might encounter when reviewing a commercial plat of survey. Other matters include, e.g., riparian rights, gaps and overlaps, and boundary concerns relative to the determination of the center of a section of land or the center of an undedicated road. Accordingly, the real estate attorney accustomed to reviewing residential plats of survey may feel at first that the commercial survey is a potential minefield of potentially explosive issues. This really is not the case. But if the attorney encounters a problem while examining a plat of survey, he or she should contact a title officer at the local title insurance company or the surveyor who prepared the survey. Either party should be able to address the attorney's concerns.

PART II

# *Certifications and Codes of Practice*

# 4

# *Minimum Standard Detail Requirements for ALTA/ACSM Land Title Surveys*

## MINIMUM STANDARD DETAIL REQUIREMENTS FOR ALTA/ACSM LAND TITLE SURVEYS

*(Effective February 23, 2011)*

**1. Purpose** - Members of the American Land Title Association (ALTA) have specific needs, unique to title insurance matters, when asked to insure title to land without exception as to the many matters which might be discoverable from survey and inspection, and which are not evidenced by the public records.

For a survey of real property, and the plat, map or record of such survey, to be acceptable to a title insurance company for the purpose of insuring title to said real property free and clear of survey matters (except those matters disclosed by the survey and indicated on the plat or map), certain specific and pertinent information must be presented for the distinct and clear understanding between the insured, the client (if different from the insured), the title insurance company (insurer), the lender, and the surveyor professionally responsible for the survey.

In order to meet such needs, clients, insurers, insureds, and lenders are entitled to rely on surveyors to conduct surveys and prepare associated plats or maps that are of a professional quality and appropriately uniform, complete and accurate. To that end, and in the interests of the general public, the surveying profession, title insurers and abstracters, the ALTA and the National Society of Professional Surveyors, Inc. (NSPS) jointly promulgate the within details and criteria setting forth a minimum standard of performance for ALTA/ACSM Land Title Surveys. A complete 2011 ALTA/ACSM Land Title Survey includes the on-site fieldwork required under Section 5 herein, the preparation of a plat or map showing the results of the fieldwork and its relationship to record documents as required under Section 6 herein, any information in Table A herein that may have been negotiated with the client, and the certification outlined in Section 7 herein.

**2. Request for Survey** - The client shall request the survey or arrange for the survey to be requested, and shall provide a written authorization to proceed from the person or entity responsible for paying for the survey. Unless specifically authorized in writing by the insurer, the insurer shall not be responsible for any costs associated with the preparation of the survey. The request shall specify that an **"ALTA/ACSM LAND TITLE SURVEY"** is required and which of the optional items listed in Table A herein, if any, are to be incorporated. Certain properties, including, but not limited to, marinas, campgrounds, trailer parks and leased areas, may present issues outside those normally encountered on an ALTA/ACSM Land Title Survey. The scope of work related to such properties should be discussed with the client, lender and insurer, and agreed upon in writing prior to requesting the survey. The client may need to secure permission for the surveyor to enter upon the property to be surveyed, adjoining properties, or offsite easements.

**3. Surveying Standards and Standards of Care**

**A. Effective Date** - The 2011 Minimum Standard Detail Requirements for ALTA/ACSM Land Title Surveys are effective February 23, 2011. As of that date, all previous versions of the Minimum Standard Detail Requirements for ALTA/ACSM Land Title Surveys are superseded by these standards.

**B. Other Requirements and Standards of Practice** - Some Federal agencies, many states and some local jurisdictions have adopted statutes, administrative rules and/or ordinances that set out standards regulating the practice of surveying within their jurisdictions. In addition to the standards set forth herein, surveyors shall also conduct their surveys in accordance with all applicable jurisdictional requirements and standards of practice. Where conflicts between the standards set forth herein and any such jurisdictional requirements and standards of practice occur, the more stringent shall apply.

**C. The Normal Standard of Care** - Surveyors should recognize that there may be unwritten local, state, and/or regional standards of care defined by the practice of the 'prudent surveyor' in those locales.

**D. Boundary Resolution** - The boundary lines and corners of any property being surveyed as part of an ALTA/ACSM Land Title Survey shall be established and/or retraced in accordance with appropriate boundary law principles governed by the set of facts and evidence found in the course of performing the research and survey.

**E. Measurement Standards** - The following measurement standards address Relative Positional Precision for the monuments or witnesses marking the corners of the surveyed property.

i. "Relative Positional Precision" means the length of the semi-major axis, expressed in feet or meters, of the error ellipse representing the uncertainty due to random errors in measurements in the location of the monument, or witness, marking any corner of the surveyed property relative to the monument, or witness, marking any other corner of the surveyed property at the 95 percent confidence level (two standard deviations). Relative Positional Precision is estimated by the results of a correctly weighted least squares adjustment of the survey.

ii. Any boundary lines and corners established or retraced may have uncertainties in location resulting from (1) the availability, condition, history and integrity of reference or controlling monuments, (2) ambiguities in the record descriptions or plats of the surveyed property or its adjoiners, (3) occupation or possession lines as they may differ from the written title lines, and (4) Relative Positional Precision. Of these four sources of uncertainty, only Relative Positional Precision is controllable, although due to the inherent errors in any measurement, it cannot be eliminated. The magnitude of the first three uncertainties can be projected based on evidence; Relative Positional Precision is estimated using statistical means (see Section 3.E.i. above and Section 3.E.v. below).

iii. The first three of these sources of uncertainty must be weighed as part of the evidence in the determination of where, in the surveyor's opinion, the boundary lines and corners of the surveyed property should be located (see Section 3.D. above). Relative Positional Precision is a measure of how precisely the surveyor is able to monument and report those positions; it is not a substitute for the application of proper boundary law principles. A boundary corner or line may have a small Relative Positional Precision because the survey measurements were precise, yet still be in the wrong position (i.e. inaccurate) if it was established or retraced using faulty or improper application of boundary law principles.

iv. For any measurement technology or procedure used on an ALTA/ACSM Land Title Survey, the surveyor shall (1) use appropriately trained personnel, (2) compensate for systematic errors, including those associated with instrument calibration, and (3) use appropriate error propagation and measurement design theory (selecting the proper instruments, geometric layouts, and field and computational procedures) to control random errors such that the maximum allowable Relative Positional Precision outlined in Section 3.E.v. below is not exceeded.

v. The maximum allowable Relative Positional

Precision for an ALTA/ACSM Land Title Survey is 2 cm (0.07 feet) plus 50 parts per million (based on the direct distance between the two corners being tested). It is recognized that in certain circumstances, the size or configuration of the surveyed property, or the relief, vegetation or improvements on the surveyed property will result in survey measurements for which the maximum allowable Relative Positional Precision may be exceeded. If the maximum allowable Relative Positional Precision is exceeded, the surveyor shall note the reason as explained in Section 6.B.ix below.

**4.** **Records Research** - It is recognized that for the performance of an ALTA/ACSM Land Title Survey, the surveyor will be provided with appropriate data which can be relied upon in the preparation of the survey. The request for an ALTA/ACSM Land Title Survey shall set forth the current record description of the property to be surveyed or, in the case of an original survey, the current record description of the parent parcel that contains the property to be surveyed. Complete copies of the most recent title commitment, the current record description of the property to be surveyed (or, in the case of an original survey, the parent parcel), the current record descriptions of adjoiners, any record easements benefiting the property, the record easements or servitudes and covenants burdening the property (all hereinafter referred to collectively as "Record Documents"), documents of record referred to in the Record Documents, documents necessary to ascertain, if possible, the junior/senior relationship pursuant to Section 6.B.vii. below, and any other documents containing desired appropriate information affecting the property being surveyed, and to which the ALTA/ACSM Land Title Survey shall make reference, shall be provided to the surveyor for use in conducting the survey. Reference is made to Section 3.B. above.

**5.** **Field Work** - The Survey shall be performed on the ground (except as otherwise negotiated pursuant to Table A, Item 15 below, if selected by the client), and the field work shall include the following:

**A. Monuments**

- **i.** The location and description of any monuments or lines that control the boundaries of the surveyed property.
- **ii.** The location, size and type of any monuments found (or set, if Table A, Item 1 is requested by the client, or if otherwise required – see Section 3.B. above) on the boundary of the surveyed property.

**B. Rights of Way and Access**

- **i.** The distance from the appropriate corner or corners of the surveyed property to the nearest right of way line, if the surveyed property does not abut a right of way.
- **ii.** The name of any street, highway or other public or private way abutting the surveyed property, and the width and location of the travelled way relative to the nearest boundary line of the surveyed property.
- **iii.** Visible evidence of physical access (such as, but not limited to, curb cuts and driveways) to any abutting streets, highways or other public ways.
- **iv.** The location and character of vehicular, pedestrian or other forms of access by other than the apparent occupants of the surveyed property to or across the surveyed property, including, but not limited to driveways, alleys, private roads, sidewalks and footpaths observed in the process of conducting the survey.
- **v.** Without expressing a legal opinion as to ownership or nature, the location and extent of any potentially encroaching driveways, alleys, and other ways of access from adjoining properties onto the surveyed property observed in the process of conducting the survey.

- **vi.** Where documentation of the width or location of any abutting street, road or highway right of way was not disclosed in Record Documents provided to the surveyor or was not otherwise available from the controlling jurisdiction (see Section 6.C.iv. below), the evidence and location of parcel corners recovered which might indicate the width or location of such right of way lines.
- **vii.** Evidence of access to and from waters adjoining the surveyed property, such as paths, boat slips, launches, piers and docks observed in the process of conducting the survey.

**C. Lines of Possession, and Improvements along the Boundaries**

- **i.** The character and location of evidence of possession or occupation along the perimeter of the surveyed property, both by the occupants of the surveyed property and by adjoiners, observed in the process of conducting the survey.
- **ii.** The character and location of all walls, buildings, fences, and other improvements within five feet of each side of the boundary lines, observed in the process of conducting the survey.
- **iii.** Without expressing a legal opinion as to the ownership or nature of the potential encroachment, the evidence, location and extent of potentially encroaching structural appurtenances and projections observed in the process of conducting the survey, such as fire escapes, bay windows, windows and doors that open out, flue pipes, stoops, eaves, cornices, areaways, steps, trim, etc., by or onto adjoining property, or onto rights of way, easements or setback lines disclosed in Record Documents provided to the surveyor.

**D. Buildings**

Based on the normal standard of care, the location of all buildings on the surveyed property shown perpendicular to the nearest perimeter boundary line(s) and expressed to the appropriate degree of precision.

**E. Easements and Servitudes**

- **i.** Evidence of any easements or servitudes burdening the surveyed property, disclosed in the Record Documents provided to the surveyor and observed in the process of conducting the survey.
- **ii.** Evidence of easements or servitudes not disclosed in the Record Documents provided to the surveyor, but observed in the process of conducting the survey, such as those created by roads; rights of way; water courses; ditches; drains; telephone, fiber optic lines, or electric lines; water, sewer, oil or gas pipelines on or across the surveyed property and on adjoining properties if they appear to affect the surveyed property.
- **iii.** Surface indications of underground easements or servitudes on or across the surveyed property observed in the process of conducting the survey.
- **iv.** Evidence of use of the surveyed property by other than the apparent occupants observed in the process of conducting the survey.

**F. Cemeteries**

As accurately as the evidence permits, the location of cemeteries, gravesites, and burial grounds (i) disclosed in the Record Documents provided to the surveyor, or (ii) observed in the process of conducting the survey.

**G. Water Features**

- **i.** The location of springs, together with the location of ponds, lakes, streams, and rivers bordering on or running through the surveyed property, observed during the process of conducting the survey. See Table A, Item 19 for wetlands locations.

ii. The location of any water boundary on the surveyed property. The attribute(s) of the water feature located (e.g. top of bank, edge of water, high water mark, etc.) should be congruent with the boundary as described in the record description or, in the case of an original survey, in the new description. (See Section 6.B.vi. below).

6. **Plat or Map** - A plat or map of an ALTA/ACSM Land Title Survey shall show the following information. Where dimensioning is appropriate, dimensions shall be in accordance with the appropriate standard of care.

A. **The evidence and locations gathered during the field work as outlined in Section 5 above.**

B. **Boundary, Descriptions, Dimensions and Closures**

i. The current record description of the surveyed property, and any new description of the surveyed property that was prepared in conjunction with the survey, including a statement explaining why the new description was prepared. Preparation of a new description should be avoided unless deemed necessary or appropriate by the surveyor and insurer. Preparation of a new description should also generally be avoided when the record description is a lot or block in a platted, recorded subdivision.

ii. The location and description of any monuments, lines or other evidence that control the boundaries of the surveyed property or that were otherwise relied upon in establishing or retracing the boundaries of the surveyed property, and the relationship of that evidence to the surveyed boundary. In some cases, this will require notes on the plat or map.

iii. All distances and directions identified in the record description of the surveyed property (and in the new description, if one was prepared). Where a measured or calculated dimension differs from the record by an amount deemed significant by the surveyor, such dimension shall be shown in addition to, and differentiated from, the corresponding record dimension.

iv. The directional, distance and curve data necessary to compute a mathematical closure of the surveyed boundary. A note if the record description does not mathematically close. The basis of bearings and, when it differs from the record basis, the difference.

v. The remainder of any recorded lot or existing parcel, when the surveyed property is composed of only a portion of such lot or parcel, shall be graphically depicted. Such remainder does not need to be included as part of the actual survey, except to the extent necessary to locate the lines and corners of the surveyed property, and it need not be fully dimensioned or drawn at the same scale as the surveyed property.

vi. When the surveyed property includes a water boundary, a note on the face of the plat or map noting the date the boundary was measured, which attribute(s) of the water feature was/were located, and the caveat that the boundary is subject to change due to natural causes and that it may or may not represent the actual location of the limit of title. When the surveyor is aware of natural or artificial realignments or changes in such boundaries, the extent of those changes and facts shall be shown or explained.

vii. The relationship of the boundaries of the surveyed property (i.e. contiguity, gaps, or overlaps) with its adjoiners, where ascertainable from Record Documents and/or from field evidence gathered during the process of conducting the survey of the property being surveyed. If the surveyed property is composed of multiple parcels, the extent of any gaps or overlaps between those parcels shall be identified. Where gaps or overlaps are identified, the surveyor shall, prior to preparation of the final plat or map, disclose this to the insurer and client for determination of a course of action concerning junior/senior rights.

**viii.** When, in the opinion of the surveyor, the results of the survey differ significantly from the record, or if a fundamental decision related to the boundary resolution is not clearly reflected on the plat or map, the surveyor shall explain this information with notes on the face of the plat or map.

**ix.** A note on the face of the plat or map explaining the site conditions that resulted in a Relative Positional Precision that exceeds the maximum allowed under Section 3.E.v. of these standards.

**x.** A note on the face of the plat or map identifying the title commitment/policy number, effective date and name of the insurer for any title work provided to the surveyor.

**C. Easements, Servitudes, Rights of Way, Access and Record Documents**

**i.** The width and recording information of all plottable rights of way, easements and servitudes burdening and benefitting the property surveyed, as evidenced by Record Documents which have been provided to the surveyor.

**ii.** A note regarding any right of way, easement or servitude evidenced by a Record Document which has been provided to the surveyor (a) the location of which cannot be determined from the record document, or (b) of which there was no observed evidence at the time of the survey, or (c) that is a blanket easement, or (d) that is not on, or does not touch, the surveyed property, or (e) that limits access to an otherwise abutting right of way, or (f) in cases where the surveyed property is composed of multiple parcels, which of such parcels the various rights of way, easements, and servitudes cross.

**iii.** A note if no physical access to a public way was observed in the process of conducting the survey.

**iv.** The width of abutting rights of way and the source of such information (a) where available from the controlling jurisdiction or (b) where disclosed in Record Documents provided to the surveyor.

**v.** The identifying titles of all recorded plats, filed maps, right of way maps, or similar documents which the survey represents, wholly or in part, with their recording or filing data.

**vi.** For non-platted adjoining land, names and recording data identifying adjoining owners according to current public records. For platted adjoining land, the recording data of the subdivision plat.

**vii.** Platted setback or building restriction lines which appear on recorded subdivision plats or which were disclosed in Record Documents provided to the surveyor.

**D. Presentation**

**i.** The plat or map shall be drawn on a sheet of not less than 8 ½ by 11 inches in size at a legible, standard engineering scale, with that scale clearly indicated in words or numbers and with a graphic scale. When recordation or filing of a plat or map is required by law, such plat or map shall be produced in recordable form. The boundary of the surveyed property drawn in a manner that distinguishes it from other lines on the plat or map. A north arrow (with north to the top of the drawing when practicable), a legend of symbols and abbreviations, and a vicinity map showing the property in reference to nearby highway(s) or major street intersection(s).

**ii.** Supplementary or detail diagrams when necessary.

**iii.** If there are no visible buildings on the surveyed property, a note stating *"No buildings existing on the surveyed property"* shall appear on the face on the survey.

**iv.** The surveyor's project number (if any), and the name, registration or license number, signature, seal, street address, telephone number, and email address of the surveyor who performed the survey. The date(s) of any revisions made by said surveyor.
**v.** Sheet numbers where the plat or map is composed of more than one sheet.
**vi.** The caption "ALTA/ACSM Land Title Survey."

**7.** **Certification** - The plat or map of an ALTA/ACSM Land Title Survey shall bear only the following certification, unaltered, except as may be required pursuant to Section 3.B. above:

*To (name of insured, if known), (name of lender, if known), (name of insurer, if known), (names of others as negotiated with the client):*

*This is to certify that this map or plat and the survey on which it is based were made in accordance with the 2011 Minimum Standard Detail Requirements for ALTA/ACSM Land Title Surveys, jointly established and adopted by ALTA and NSPS, and includes Items _____ of Table A thereof. The field work was completed on ___________.*

Date of Plat or Map:_____ (Surveyor's signature, printed name and seal with Registration/License Number)

**8.** **Deliverables** - The surveyor shall furnish copies of the plat or map of survey to the insurer and client, and as otherwise negotiated with the client. Hard copies shall be on durable and dimensionally stable material of a quality standard acceptable to the insurer. Digital copies of the plat or map may be provided in addition to, or in lieu of, hard copies in accordance with the terms of the contract. When required by law or requested by the client, the plat or map shall be produced in recordable form and recorded or filed in the appropriate office or with the appropriate agency.

## *TABLE A*

### *OPTIONAL SURVEY RESPONSIBILITIES AND SPECIFICATIONS*

*NOTE: The items of Table A must be negotiated between the surveyor and client. It may be necessary for the surveyor to qualify or expand upon the description of these items (e.g., in reference to Item 6(b), there may be a need for an interpretation of a restriction). The surveyor cannot make a certification on the basis of an interpretation or opinion of another party. Notwithstanding Table A Items 5 and 11(b), if an engineering design survey is desired as part of an ALTA/ACSM Land Title Survey, such services should be negotiated under Table A, item 22.*

***If checked, the following optional items are to be included in the ALTA/ACSM LAND TITLE SURVEY, except as otherwise qualified (see note above):***

1. _____ *Monuments placed (or a reference monument or witness to the corner) at all major corners of the boundary of the property, unless already marked or referenced by existing monuments or witnesses.*

2. _____ *Address(es) if disclosed in Record Documents, or observed while conducting the survey.*

3. _____ *Flood zone classification (with proper annotation based on federal Flood Insurance Rate Maps or the state or local equivalent) depicted by scaled map location and graphic plotting only.*

4. _____ *Gross land area (and other areas if specified by the client).*

5. _____ *Vertical relief with the source of information (e.g. ground survey or aerial map), contour interval, datum, and originating benchmark identified.*

6. _____ *(a) Current zoning classification, as provided by the insurer.*

   _____ *(b) Current zoning classification and building setback requirements, height and floor space area restrictions as set forth in that classification, as provided by the insurer. If none, so state.*

7. _____ *(a) Exterior dimensions of all buildings at ground level.*

   *(b) Square footage of:*

   _____ *(1) exterior footprint of all buildings at ground level.*

   _____ *(2) other areas as specified by the client.*

   _____ *(c) Measured height of all buildings above grade at a location specified by the client. If no location is specified, the point of measurement shall be identified.*

8. _____ *Substantial features observed in the process of conducting the survey (in addition to the improvements and features required under Section 5 above) such as parking lots, billboards, signs, swimming pools, landscaped areas, etc.*

9. _____ *Striping, number and type (e.g. handicapped, motorcycle, regular, etc.) of parking spaces in parking areas, lots and structures.*

10. _____ *(a) Determination of the relationship and location of certain division or party walls designated by the client with respect to adjoining properties (client to obtain necessary permissions).*

_____ *(b) Determination of whether certain walls designated by the client are plumb (client to obtain necessary permissions).*

11. *Location of utilities (representative examples of which are listed below) existing on or serving the surveyed property as determined by:*

_____ *(a) Observed evidence.*

_____ *(b) Observed evidence together with evidence from plans obtained from utility companies or provided by client, and markings by utility companies and other appropriate sources (with reference as to the source of information).*

- *Railroad tracks, spurs and sidings;*
- *Manholes, catch basins, valve vaults and other surface indications of subterranean uses;*
- *Wires and cables (including their function, if readily identifiable) crossing the surveyed property, and all poles on or within ten feet of the surveyed property. Without expressing a legal opinion as to the ownership or nature of the potential encroachment, the dimensions of all encroaching utility pole crossmembers or overhangs; and*
- *utility company installations on the surveyed property.*

*Note - With regard to Table A, item 11(b), source information from plans and markings will be combined with observed evidence of utilities to develop a view of those underground utilities. However, lacking excavation, the exact location of underground features cannot be accurately, completely and reliably depicted. Where additional or more detailed information is required, the client is advised that excavation may be necessary.*

12. _____ *Governmental Agency survey-related requirements as specified by the client, such as for HUD surveys, and surveys for leases on Bureau of Land Management managed lands.*

13. _____ *Names of adjoining owners of platted lands according to current public records.*

14. _____ *Distance to the nearest intersecting street as specified by the client.*

15. _____ *Rectified orthophotography, photogrammetric mapping, airborne/mobile laser scanning and other similar products, tools or technologies as the basis for the*

showing the location of certain features (excluding boundaries) where ground measurements are not otherwise necessary to locate those features to an appropriate and acceptable accuracy relative to a nearby boundary. The surveyor shall (a) discuss the ramifications of such methodologies (e.g. the potential precision and completeness of the data gathered thereby) with the insurer, lender and client prior to the performance of the survey and, (b) place a note on the face of the survey explaining the source, date, precision and other relevant qualifications of any such data.

16. _____ Observed evidence of current earth moving work, building construction or building additions.

17. _____ Proposed changes in street right of way lines, if information is available from the controlling jurisdiction. Observed evidence of recent street or sidewalk construction or repairs.

18. _____ Observed evidence of site use as a solid waste dump, sump or sanitary landfill.

19. _____ Location of wetland areas as delineated by appropriate authorities.

20. _____ (a) Locate improvements within any offsite easements or servitudes benefitting the surveyed property that are disclosed in the Record Documents provided to the surveyor and that are observed in the process of conducting the survey (client to obtain necessary permissions).

_____ (b) Monuments placed (or a reference monument or witness to the corner) at all major corners of any offsite easements or servitudes benefitting the surveyed property and disclosed in Record Documents provided to the surveyor (client to obtain necessary permissions).

21. _____ Professional Liability Insurance policy obtained by the surveyor in the minimum amount of $___________ to be in effect throughout the contract term. Certificate of Insurance to be furnished upon request.

22. _____ ______________________________________________

Adopted by the Board of Governors, American Land Title Association, on October 13, 2010.
American Land Title Association, 1828 L St., N.W., Suite 705, Washington, D.C. 20036.

Adopted by the Board of Directors, National Society of Professional Surveyors, on November 15, 2010.
National Society of Professional Surveyors, Inc., a member organization of the American Congress on Surveying and Mapping, 6 Montgomery Village Avenue, Suite 403, Gaithersburg, MD 20879

# The 2011 ALTA/ACSM Land Title Survey Standards[1]

5

Richard F. Bales and Gary R. Kent

## I. Introduction

The 2011 Minimum Standard Detail Requirements for ALTA/ACSM Land Title Surveys [hereinafter 2011 Standards] went into effect as of February 23, 2011—the date of the festival held in honor of Terminus, the Roman god who protected boundary markers. These new standards represent the first major revision of the land title survey standards since they were adopted in 1962 by the American Congress on Surveying and Mapping (ACSM) and the American Title Association (now the American Land Title Association, or ALTA). These new standards were drafted by an ALTA/ACSM liaison committee chaired by Gary R. Kent, an Indianapolis surveyor and one of the authors of this article. These new standards can be found at www.alta.org/forms in the "Most Requested" section. This article will discuss the most significant changes to these standards.

1. An earlier version of this article appeared in PROBATE & PROPERTY (Nov./Dec. 2011).

## II. Changes to the Main Text of the Standards

In general, the 2011 Standards have been completely reformatted and reorganized. The 2005 ALTA/ACSM land title survey standards [hereinafter 2005 Standards] contained a few untitled preliminary paragraphs, followed by an extensive section that detailed what the plat of survey should include.

The 2011 Standards contain eight numbered sections. These standards begin with a "Purpose" section that sets out the premise that title companies, lenders, and their respective customers are entitled to rely on plats of survey that are appropriately uniform, complete, and accurate. This section also includes a definition of a "2011 ALTA/ACSM Land Title Survey":

> A complete 2011 ALTA/ACSM Land Title Survey includes the on-site fieldwork required under Section 5 herein, the preparation of a plat or map showing the results of the fieldwork and its relationship to record documents as required under Section 6 herein, any information in Table A herein that may have been negotiated with the client, and the certification outlined in Section 7 herein.

This paragraph illustrates a fundamental change in the structure of the survey standards. As noted above, the 2005 Standards contained a lengthy section (section 5) that sets forth what the plat of the land title survey should contain. But it contains a mixture of other things as well, such as fieldwork requirements and calculations. Thus, the ALTA/ACSM liaison committee decided that it would make more sense to have one section that detailed what had to be done in the field to assure that everything the title company needed to issue its title policy(s) was gathered by the surveyor. This section was followed by a section that addressed—now that the fieldwork was done—the transfer of the survey fieldwork onto the plat (or map) of survey.

Section 2, entitled "Request for Survey," provides that the surveyor's client should request the survey and that the request should include which, if any, of the optional items listed in Table A should be incorporated into the plat of survey. This section also includes some guidance for surveying nonstandard real estate:

> Certain properties, including, but not limited to, marinas, campgrounds, trailer parks and leased areas, may present issues outside those normally encountered on an ALTA/ACSM Land Title Survey. The scope of work related to such properties should be discussed with the client, lender and insurer, and agreed upon in writing prior to requesting the survey. The client may need to secure permission for the surveyor to enter upon the property to be surveyed, adjoining properties, or offsite easements.

Note the naming of four specific properties: marinas, campgrounds, trailer parks, and leased areas. Why were these four property types singled out? They were noted in this section because Gary Kent routinely receives phone calls and e-mails from surveyors asking about these types of properties. For example, with marinas, surveyors are concerned about possible riparian rights. With marinas, campgrounds, and trailer parks, surveyors often ask, “What do I locate and show on my plat of survey?” With leased real estate, surveyors will simply wonder, “Can I survey a leased area?” Ultimately, this paragraph was drafted to encourage surveyors to communicate with their clients and with title companies so that everyone has the same expectations.

Section 3 is entitled “Surveying Standards and Standards of Care.” This section illustrates how the drafters of the 2011 Standards walked the fine line between drafting a national standard while still recognizing a possible local standard of care. That is, the liaison committee wanted to create an objective survey standard. The committee wanted the surveyor to essentially follow an objective set of guidelines so that when the surveyor was finished, he could say to himself, “I did everything the standards require; therefore, I can certify that this plat of survey meets these survey standards.”

On the other hand, because these standards are national standards but are used on a local level, the liaison committee felt compelled to mention (1) that there may be written statutes, administrative rules, and/or ordinances that regulate the practice of surveying, and (2) that there may be subjective unwritten local, state, and/or regional standards of care as defined and practiced by the “prudent surveyor” in a particular geographical location.

Thus, the committee had to deal with three sets of survey guidelines: (1) the actual ALTA/ACSM land title survey standards; (2) written statutes, administrative rules, and/or ordinances; and (3) an unwritten and therefore subjective and possibly nebulous standard of care for the “prudent surveyor” in the community or regional area.

The problem confronting the ALTA/ACSM liaison committee (and its solution) is set forth as follows:

> Problem: The committee wanted to create an objective land title survey standard, but it also wanted to remind the surveyor to be aware of a possible subjective local or regional standard of care.
>
> Solution: The 2011 Standards require surveyors to “conduct their surveys” in accordance with both the ALTA/ACSM land title survey standards and the written statutes, administrative rules, and/or ordinances. But the standards also require that the surveyor “recognize that there may be unwritten local, state, and/or regional standards of care defined by the practice of the ‘prudent surveyor’ in those locales.” (emphasis added)

In the past, these survey standards have not addressed the issue of properly establishing the boundary of the surveyed property. Section 3.D now provides for this:

> The boundary lines and corners of any property being surveyed as part of an ALTA/ACSM Land Title Survey shall be established and/or retraced in accordance with appropriate boundary law principles governed by the set of facts and evidence found in the course of performing the research and survey.

Section 3.E modifies and expands on the measurement precision standard of the land title survey. Before, the term of art was Relative Positional Accuracy. Now, the term is Relative Positional Precision. The committee is currently working to develop a document that would not be part of the 2011 Standards, but instead would be a reference for surveyors with respect to this measurement standard.

Section 4 is entitled "Records Research." It provides that the so-called "Record Documents" be furnished the surveyor. These documents include the following items:

> Complete copies of the most recent title commitment, the current record description of the property to be surveyed (or, in the case of an original survey, the parent parcel), the current record descriptions of adjoiners, any record easements benefiting the property, the record easements or servitudes and covenants burdening the property.

Contrary to popular belief, this section does not require that the title company perform a title search of the adjoining property. Rather, the surveyor needs only the "current record descriptions of adjoiners." However, this section 4 requirement should be read in conjunction with paragraph 6.C.vi of the 2011 Standards, which requires that the plat of survey show "for non-platted adjoining land, names and recording data identifying adjoining owners according to current public records. For platted adjoining land, the recording data of the subdivision plat." Thus, the "Record Documents" relating to adjoining land will be copies of current recorded deeds and, if appropriate, either one or more subdivision plats or the recording data of said plat(s).

But who should furnish these documents to the surveyor? Although ALTA anticipated that any records research would probably be performed by the title insurance company, the 2011 Standards do not state that the title company is responsible for providing these documents. Rather, section 4 of these standards merely indicates that this information "shall be provided to the surveyor for use in conducting the survey." The standards do not specify who shall obtain these documents and furnish them to the surveyor.

However, if Client makes a contract with Surveyor to survey Blackacre and to prepare a land title survey according to the 2011 Standards, then these standards would essentially be part of this contract between Surveyor and Client. Contractually, then, one might argue that the client should be furnishing the surveyor these "Record Documents." Practically, though, in many instances the title company will be the logical source for these documents.

Notwithstanding the dictates and fine points of section 4, note that in many states surveyors are required by administrative rule to conduct their own deed research. A surveyor in those states may never ask a title company for copies of recorded deeds, plats, and other documents unless he encounters difficulties while conducting his title search. Additionally, there is nothing in the standards that requires such documents be provided at no cost.

Section 5, entitled, "Fieldwork," is the heart of the standards. This section provides that "the fieldwork shall include the following [subjects to be addressed by the surveyor]": monuments; rights-of-way and access; lines of possession and improvements along the boundaries; buildings; easements and servitudes; cemeteries; and water features.

This section contains several new elements. For example, the issue of access (paragraph 5.B.iv.) is greatly expanded:

> The location and character of vehicular, pedestrian or other forms of access by other than the apparent occupants of the surveyed property to or across the surveyed property, including, but not limited to driveways, alleys, private roads, sidewalks and footpaths observed in the process of conducting the survey.

Section 5.D requires the surveyor to show the location of all buildings on the surveyed property "based on the normal standard of care." That is, what is the surveyor's standard of care relative to his surveying a particular property? For example, if a surveyor is surveying an office building in downtown Chicago, can he describe the building as being "on line" if it is located to the nearest 0.01 foot? On the other hand, if a surveyor has to locate a barn in the middle of a quarter section (2,640 feet square) that is in a rural countryside, is it appropriate if he locates said barn on his plat of survey to the nearest 10 feet? Depending on where the surveyor is working and the type of property he is surveying, the standard of care may be different. The ALTA/ACSM liaison committee wanted to emphasize that surveyors had to know what the appropriate standard of care was in any given situation and that surveyors had to apply that standard of care accordingly.

Section 5.G is entitled "Water Features," and paragraph 5.G.ii is a new water rights paragraph:

> The location of any water boundary on the surveyed property. The attribute(s) of the water feature located (e.g., top of bank, edge of water, high water mark, etc.) should be congruent with the boundary as described in the record description or, in the case of an original survey, in the new description.

For example, if a legal description contains a call, "to the west bank of the creek," then the surveyor should locate and show the west bank of the creek on his plat of survey. Similarly, if there is a call in a legal description that reads "to the edge of the river," the surveyor should locate and show the edge of the river on his plat of survey. In other words, the committee felt that there should be some connection, some agreement, as to how the legal description describes the boundary of the surveyed property and what the surveyor actually locates and shows on his plat of survey.

This would be the case, regardless of whether the surveyor was performing a retracement survey and using an existing legal description or surveying a newly created tract of land and writing a brand new legal description. Thus, if the surveyor is performing a retracement survey of an existing parcel of land and the legal description contains the call, "to the west bank of the Fox River," he should show the west bank of the river on his plat of survey.

On the other hand, consider this example: a surveyor is conducting an original survey for what will become the "first deed out" of a forty-four-acre tract of land carved out of a quarter section of land. Blackberry Creek flows through this quarter section. The surveyor's client asks that the center of Blackberry Creek be the eastern boundary of the new tract of land being surveyed. Accordingly, because the client has requested that the surveyor's legal description of this forty-four-acre tract reference the center of Blackberry Creek, the surveyor's plat of survey should similarly locate the center of the creek and represent it as the eastern boundary of this newly created parcel of land.

Other water-related paragraphs in the 2011 Standards are 5.B.vii and 6.B.vi.

Section 6 is entitled "Plat or Map," and this section provides that "a plat or map of an ALTA/ACSM Land Title Survey shall show the following information. Where dimensioning is appropriate, dimensions shall be in accordance with the appropriate standard of care." Note that the dimensioning of features shown on a plat of survey has never been addressed in previous versions of the standards, except with regard to building locations, and that was in the 2005 Standards.

Section 6.A carries forward the surveyor's fieldwork described in section 5. That is, this section provides that the plat of survey shall show "the evidence and locations gathered during the fieldwork as outlined in Section 5 above."

Section 6.B concerns "Boundary, Descriptions, Dimensions and Closures." Paragraph 6.B.i provides that the "preparation of a new [legal] description should be avoided unless deemed necessary or appropriate by the surveyor and insurer." Surveyors will sometimes create a new legal description of the land being surveyed. They may not like the existing legal description, or the proposed lender may ask that the surveyor write a new description. Oftentimes, however, there is no real need for a new legal description. This paragraph allows, but discourages, the unnecessary rewriting of legal descriptions.

Paragraph 6.B.v clarifies how the remainder of a parcel should be depicted on a plat of survey, noting that "such remainder does not need to be included as part of the actual survey, except to the extent necessary to locate the lines and corners of the surveyed property . . . "

Paragraph 6.B.vii concerns the "gap/overlap" or "junior and senior rights" issue. This paragraph states that "where gaps or overlaps are identified, the surveyor shall, prior to preparation of the final plat or map, disclose this to the insurer and client for determination of a course of action concerning junior/senior rights." On a national level, surveyors are not in agreement as to how they should address the issue of junior and senior rights. In some parts of the country, surveyors address gaps and overlaps by attempting to resolve the junior and senior relationships during the process of conducting their survey of the land. But in most parts of the country, surveyors simply identify and disclose gaps and overlaps between current record descriptions on their plats of survey. The ALTA/ACSM liaison committee had this latter procedure in mind when it drafted paragraph 6.B.vii. That is, the committee viewed the junior and senior rights problem as a title issue to be addressed collaboratively by both the surveyor and the title company. The committee did not feel it was a survey matter to be resolved solely by the surveyor.

Paragraph 6.B.x sets forth the requirement that the title commitment or policy information (commitment or policy number, commitment or policy effective date, and name of title insurance company) of any title work furnished to the surveyor be shown on the plat of survey.

Section 6.C is entitled "Easements, Servitudes, Rights-of-Way, Access and Record Documents." Paragraph 6.C.i requires that the width of all rights-of-way, easements, and servitudes be shown on the survey. ("Width" was not specified in corresponding paragraph 5(h) of the 2005 Standards.) Paragraph 5(h) of the 2005 Standards required that if an easement could not be located on the plat of survey, "a note to this effect shall be included." Paragraph 6.C.ii of the 2011 Standards greatly expands upon this concept of offering explanatory notes concerning easements. Six topics are mentioned:

> A note regarding any right of way, easement or servitude evidenced by a Record Document which has been provided to the surveyor (a) the location of which cannot be determined from the record document, or (b) of which there was no observed evidence at the time of the survey, or (c) that is a blanket easement, or (d) that is not on, or does not touch, the surveyed property, or (e) that limits access to an otherwise abutting right of way, or (f) in cases where the surveyed property is composed of multiple parcels, which of such parcels the various rights of way, easements, and servitudes cross.

Section 6.D is entitled "Presentation." Paragraph 6.D.i requires, among other things, "a vicinity map showing the property in reference to nearby highway(s) or major street intersection(s)." (A vicinity map was merely an optional Table A item in the 2005 Standards.)

Section 7 is entitled "Certification." For years, surveyors have struggled with lenders who demand "long form certifications" on their land title surveys. These certifications invariably require the surveyor to make express guarantees and warranties that expose the surveyor to excessive and unreasonable liability. Additionally, the language used in such certifications often results in the surveyors possibly violating the terms of their state-issued licenses. For example, a surveyor who certifies that there are "no violations of zoning setbacks" or that there are "no encroachments" may be making a legal determination.

This section makes it clear that the plat of survey shall bear only the certification set forth in this section, unaltered except as may be required by statutes, administrative rules, or ordinances adopted by the federal agencies, states, and local jurisdictions mentioned in section 3.B of the 2011 Standards. (Note that state surveying licensing boards sometimes require that certain wording be included in the surveyor's certificate. The inclusion of such wording in the surveyor's certificate would be a permitted exception to the generally unalterable certificate language set forth in section 7.)

On the other hand, if a surveyor and a lender wish to contract for the execution of an alternative certification, they can do so, but this nonstandard certification will have to be provided on a separate document (e.g., in a letter that references the plat of survey). A hybrid certificate cannot be placed on the actual plat.

In an important and notable development, the U.S. Department of Housing and Urban Development (HUD) has already issued new HUD Lean 232 Survey Instructions that specify in paragraph I.G. that the certification in section 7 of the ALTA/ACSM Standards be used for HUD surveys. It is hoped that this will encourage lenders to take the same step.

The surveyor has to include the date of the plat of survey in this certification, but pursuant to the 2011 Standards, he also has to add the date the fieldwork was completed. This dual-date dichotomy may cause lenders some concern. Lenders' closing or escrow instructions usually include a requirement that they be furnished a "current" plat of survey. A "current" plat of survey is usually defined as a plat that is six months old or less. But how is this six-month time period determined? Is it measured from the date of the plat of survey or from the date the fieldwork was completed?

The lender requirement of a "current" plat of survey stems from the reasoning that the older the plat of survey, the greater the chance the plat will not be an accurate depiction of all physical improvements located on the land. Therefore, it seems likely that lenders will insist that the six months be measured from the date the fieldwork was completed and not the (usually later) date that the surveyor signs the completed plat of survey.

Section 8 is entitled "Deliverables." Bowing to the Internet Age, the drafters of the 2011 Standards added a sentence indicating that "digital copies of the plat or map may be provided in addition to, or in lieu of, hard copies in accordance with the terms of the contract." Note, though, that in some states, the plat of survey must have an embossed seal and an original signature in order for the plat to be valid. In other states, this is not a requirement, but surveyors will often include a statement on their plat indicating, e.g., that "this plat is not valid without the surveyor's original seal." Surveyors will do this to protect themselves from the unauthorized use of their surveys. Keep in mind, however, that a digital copy of such a plat would not be a valid survey.

## III. Changes to Table A

Table A is a listing of "Optional Survey Responsibilities and Specifications." This table has also been revised by the red pen of the ALTA/ACSM liaison committee. The introduction to Table A in the 2005 Standards included a statement about surveying for the U.S. Department of Housing and Urban Development. That statement has been eliminated from the 2011 Standards. (However, see item 12 of Table A of the new standards, which is described below.)

The committee has included a new statement in the introduction to the 2011 Standards. It provides that "notwithstanding Table A items 5 and 11(b), if an engineering design survey is desired as part of an ALTA/ACSM Land Title Survey, such services should be negotiated under Table A, item 22."

Table A, item 2, is a new requirement: "Address(es) if disclosed in Record Documents, or observed while conducting the survey." (As noted above, the former Item 2, a vicinity map, is now a requirement in Paragraph 6.D.i.)

Table A, item 5, concerning "vertical relief" and "contour interval," has been expanded and clarified.

Table A, item 6, concerns zoning. In the 2011 standards, this item has been expanded into two Table A paragraphs, 6(a) and 6(b):

> 6(a): Current zoning classification, as provided by the insurer.
>
> 6(b): Current zoning classification and building setback requirements, height and floor space area restrictions as set forth in that classification, as provided by the insurer. If none, so state.

Note that in both paragraphs 6(a) and 6(b), the zoning information is to be provided by the title insurer. If the surveyor obtained these zoning materials from a person or entity other than the title company, would that result in the survey not being an ALTA/ACSM land title survey? No, it would not. This is made clear in the introductory paragraph to Table A:

> Note: The items of Table A must be negotiated between the surveyor and client. It may be necessary for the surveyor to qualify or expand upon the description of these items (e.g., in reference to Item 6(b), there may be a need for an interpretation of a restriction).

In Mr. Kent's experience, surveyors are not usually asked to add zoning information on their plats of surveys unless the title company is including some kind of zoning endorsement with its title policy(s). Thus, the ALTA/ACSM liaison committee felt that the title company could provide to the surveyor preparing the survey a copy of the zoning materials that it used in drafting its zoning endorsement. However, feeling the heat of potential liability that is unfettered by title policy exclusions and conditions, it is possible that a title company might be reluctant to furnish zoning information to a surveyor. The liaison committee recognized this possibility, and that is why items 6(a) and 6(b) do not prohibit some other third party (or even the surveyor) from obtaining this zoning information. That is, the surveyor could acquire the zoning information from another source and still sign the survey certification, as set forth in paragraph 7 of the 2011 Standards. (If the surveyor wanted, he could even sign the plat of survey but indicate, e.g., that item 6(a) or 6(b) was "as negotiated.")

Title insurance zoning coverage is highly regulated in some states and even prohibited in at least one state. Thankfully, the failure to include items 6(a) or 6(b) on a land title survey will not invalidate its "land title survey" status. Table A makes it clear that these items, like all of the items in Table A, are "optional survey responsibilities and specifications."

Table A, item 7, concerns square footage. Table A, item 7(b)(2) of the 2005 Standards was "gross floor area of all buildings." This wording has been deleted from the 2011 Standards. Depending on the type of building, the request to compute "gross floor area" pursuant to the 2005 Standards raised several questions as to what was required of the surveyor. The determination of "gross floor area" suggests that the surveyor would have to make measurements inside a building. In that event, how would the surveyor address such features as atriums, stairwells, balconies, and elevator shafts? Item 7(b)(3) of the 2005 Standards was, "other areas to be defined by the client." This item, slightly reworded, is now Table A, Item 7(b)(2) in the 2011 Standards: "other areas as specified by the client."

Table A, item 10 of the 2005 Standards was "indication of access to a public way on land such as curb cuts and driveways, and to and from waters adjoining the surveyed tract, such as boat slips, launches, piers and docks." The issue of access is crucial to title companies, and so the disclosure of access is now mandatory under the 2011 Standards. In this regard, see paragraphs 5.B.iii and 5.B.vii of these standards.

Table A, item 10(a) of the 2011 Standards concerns party walls, and Table A, item 10(b) of these new standards deals with whether certain walls are plumb. These two new items were created from the last sentence of paragraph 5(i) of the 2005 Standards. The ALTA/ACSM liaison committee felt that whether a wall is legally a "party wall" or whether a wall is "plumb" may require extra work, and any such work should be negotiated between the surveyor and the client. Therefore, the committee felt that these two issues should be moved from the main text of the standards and become an optional item in Table A. (Walls are also covered in paragraph 5.C.ii of the 2011 Standards.)

Table A, item 11, concerns the location of utilities existing on or serving the surveyed property. Due to confusion and misunderstanding as to the surveyor's ability to accurately locate underground utilities, a qualifying note has now been added to this item:

> Note: With regard to Table A, item 11(b), source information from plans and markings will be combined with observed evidence of utilities to develop a view of those underground utilities. However, lacking excavation, the exact location of underground features cannot be accurately, completely and reliably depicted. Where additional or more detailed information is required, the client is advised that excavation may be necessary.

Item 12 of Table A of the 2005 Standards was "Governmental Agency survey-related requirements as specified by the client." Item 12 has been expanded in 2011:

> 12. Governmental Agency survey-related requirements as specified by the client, such as for HUD surveys, and surveys for leases on Bureau of Land Management managed lands.

Item 15 of Table A concerns the use of alternative technologies, such as photogrammetric mapping. The 2005 Standards included the term "laser scanning." The 2011 Standards replaces this term with "airborne/mobile laser scanning," thus making it clear that laser scanning can be either terrestrial (i.e., ground-based) or airborne.

The committee made some minor changes to the wording of items 16 and 17:

> 16. Observed evidence of current earth moving work, building construction or building additions.
>
> 17. Proposed changes in street right of way lines, if information is available from the controlling jurisdiction. Observed evidence of recent street or sidewalk construction or repairs.

Item 19 is new:

> 19. Location of wetland areas as delineated by appropriate authorities.

Item 20, consisting of two lettered paragraphs, is also a new addition to the 2011 Standards:

> 20(a): Locate improvements within any offsite easements or servitudes benefiting the surveyed property that are disclosed in the Record Documents provided to the surveyor and that are observed in the process of conducting the survey (client to obtain necessary permissions).

> 20(b): Monuments placed (or a reference monument or witness to the corner) at all major corners of any offsite easements or servitudes benefiting the surveyed property and disclosed in Record Documents provided to the surveyor (client to obtain necessary permissions).

Item 20(a) should be of particular interest to both lenders and title insurers. When title companies are asked to insure easements, they (and any new lender) may want to have the easement parcel surveyed to make sure that access through the easement area is still open and not obstructed in any fashion. Item 20(a) authorizes the surveyor to provide this additional service.

Note, though, that because any offsite easement parcel would most likely not be owned by the surveyor's client, the ALTA/ACSM liaison committee included in this item the requirement that the client obtain the necessary consent to allow the surveyor onto the easement property. (A similar statement is set forth in the last sentence of section 2 of the 2011 Standards.)

Item 21 is also new:

> 21. Professional Liability insurance policy obtained by the surveyor in the minimum amount of $_____ to be in effect throughout the contract term. Certificate of insurance to be furnished upon request.

But what exactly is the contract term? Should this professional liability insurance policy (i.e., "errors and omissions" insurance policy) be in effect only through, for example, the date of the title insurance policy? Or should the contract term continue for one or more years after this date, in case a survey problem is not discovered until months and months after the closing? The answer is found in the first sentence of Table A. This item, like all of the optional items set forth in Table A, "must be negotiated between the surveyor and the client."

When surveyors are involved in these negotiations, they should keep in mind that most states have statutes that limit the length of time that surveyors are liable for causes of action relating to their professional services. For example, § 13-222 of the Illinois Code of Civil Procedure (735 ILL. COMP. STAT. 5/13-222) provides that the surveyor is not liable for any negligence, errors, omissions, torts, or breaches of contract in the making of any survey completed on or after January 1, 2002, after more than four years have elapsed from the time the person claiming damages knew or should have known of such negligence, errors, omissions, or other actionable conduct. Furthermore, any action against the surveyor cannot be brought if ten years have elapsed from the time of the act or omission. However, any person who discovers the act or omission before the expiration of the ten-year period may in no event have less than four years to bring a cause of action. If a person entitled to bring the action is under the age of 18 or under a legal disability, the period of limitation does not begin to run until the person reaches 18 years of age or until the disability is removed. (For additional information concerning surveyor liability, see the author's article, *Land Surveyor Liability to Third Parties in Illinois*, which appeared in the March 2007 issue of the ILLINOIS BAR JOURNAL.)

These statutes will help to define the "item 21 contract term" unless the parties negotiate a different time frame. Surveyors should review their insurance policies to make sure that they are not contractually agreeing to something that is outside the scope of their policies.

The amount of insurance in this Table A item is also a matter that must be negotiated by the surveyor and the client. The client must consider the answer to these questions: If the surveyor I have hired were to make a mistake in the survey of this property, what kind of mistake might it be? What would be the potential monetary damages resulting from this mistake? The answers to these questions will depend, at least in part, on the value and location of the land and the type of improvements, if any, constructed on the land. For example, it seems reasonable to conclude that the potential damages resulting from an error made while surveying a twenty-acre tract of vacant land in the middle of a rural countryside in downstate Illinois will not be as great as the potential for damages stemming from an error made while surveying a twenty-story office building in downtown Chicago.

## IV. Miscellaneous Comments

The ALTA/ACSM liaison committee attempted to standardize the wording of the 2011 Standards with regard to the property being surveyed. In the past, the land being surveyed has been referred to as the premises, the property, the parcel, and the tract. The new standards use the term "the surveyed property," except where the use of this term is not appropriate.

Also, earlier versions of these standards included such terms as visible, observed, observable, and physical. The 2011 Standards use the term "observed in the process of conducting the survey" when possible and where appropriate.

It has been suggested that this new phrase, "observed in the process of conducting the survey," provides an "out" for the surveyor. That is, if a surveyor fails to show a manhole on his plat of survey, he can simply argue that "paragraph 5.E.ii and section 6.A of the 2011 survey standards provides that I have to show on my plat of survey all evidence of easements observed in the process of conducting the survey. However, I did not observe any evidence of this easement while I was surveying this property."

Such a conclusion seems misplaced. One might make this argument with any of the other words used in previous versions of these standards. For example:

Paragraph 5(h) of the 2005 Standards: "Observable evidence of easements and/or servitudes of all kinds . . . "

Paragraph 5.E.ii of the 2011 Standards: "Evidence of easements or servitudes . . . observed in the process of conducting the survey. . . "

That is, a manhole partially hidden under a lilac bush is likely "observable," and it is equally likely to be observed in the process of conducting the survey.

Assume for the moment that a surveyor fails to note the manhole cover on his plat of survey, and he argues that the cover was not "observed in the process of conducting the survey." The resolution of this issue is not this artificial and possibly unrealistic parsing of the 2011 Standards. Rather, the answer is found in section 3.C of these standards—the aforementioned "unwritten local, state, and/or regional standards of care defined by the practice of the 'prudent surveyor' in those locales." That is, under the normal standard of care of surveyors (e.g., in northern Illinois), should the surveyor have observed the manhole cover while he was surveying the property?

## V. Conclusion

The 2011 ALTA/ACSM land title survey standards are the culmination of two years of work by a committee made up of land surveyors, attorneys who represent lenders, and title company attorneys. The committee's work was made difficult due to the inherent challenges of writing a national standard. For example, the surveyor's final product is called a "plat" in most states but a "map" in a few states. This is why the 2011 Standards refer to a "plat or map." Another issue concerned whether the setting of monuments at the corners of the surveyed property should be mandatory or optional. In some states, setting a monument would require the filing of a "Record of Survey" for jurisdictional or administrative review, and this is why such monumentation is an optional item in Table A. A third issue was the determination of when a new legal description should be prepared.

The ALTA/ACSM liaison committee feels that it has met these challenges and has drafted a new set of survey standards that will successfully meet the differing needs and requirements of surveyors, lenders, and title companies in 2011 and the coming years.

# The Annotated Surveyor's Certificate

# 6

Mitchell G. Williams

The debates over survey certifications and their scope seem never-ending. This article addresses a portion of that debate: the notorious (at least among surveyors) "insurance company" or "long form" survey certification. The article's first goal is to point out particular problems with the "long form," especially those items that are not within the land surveyor's area of expertise or that are more appropriately determined by another professional. The article's second (and perhaps more important) goal is to show that all of the appropriate items contained in the "long form" (as well as many items not addressed in the "long form") are contained in the 2011 "Minimum Standard Detail Requirements for ALTA/ACSM Land Title Surveys" (the Code), and to show that attorneys and surveyors can adopt the Code and its certification as the "standard" for all title surveys.

Following is a typical example of an "insurance company" or "long form" certification. Throughout the certification form, the article contains notes and comments on the form's various provisions, along with occasional criticism of some of those provisions.

> The undersigned hereby certifies to ____[Owner]_______, __[Lender]____, and _________________Title Insurance Company ("Title Company") and their respective successors and assigns that:

(i) the survey dated _______________, 20__ and prepared by the undersigned (the "Survey"), is an accurate survey of the premises shown hereon and described more fully hereafter (the "Premises");

*Note:* This statement is really a "warranty" that the survey is accurate. The ALTA/ACSM certification certifies that the survey was done in accordance with the Code, which should be sufficient. The determination of boundary lines is an opinion—a professional, informed opinion, but an opinion nonetheless—and all measurements are made to a certain level of precision and contain some uncertainty. A surveyor is not a guarantor. A certificate can be to the best of the surveyor's knowledge or in the exercise of his or her professional judgment, or words to similar effect. A flat-out warranty may also void or limit the surveyor's malpractice insurance.

(ii) this Survey was made on the ground and is based on an inspection of the Premises, and it shows the location of all improvements, structures, buildings, shorelines, wetlands, streams, ponds, cemeteries, burial grounds, roads, trails, paths, walks, railroad tracks, spurs and sidings, and all other physical conditions which are either existing on the Premises or presently being constructed on the Premises;

*Note*: This language includes the location of underground improvements. A surveyor cannot certify to things that by definition are not visible. Surface indications of underground matters, e.g., manholes, are covered by other requirements of the certification (see xiv below). It is appropriate for the surveyor to limit the scope of the inspection to a surface inspection. If there are specific concerns with some suspected or known underground encroachments or unrecorded easements, the attorney should deal with them on a case-by-case basis in consultation with the surveyor.

"Wetlands" designations are highly technical and subject to the interpretation of complex and rapidly changing regulations. Thus, absent some special qualifications, surveyors would not be the professionals of choice to make wetlands determinations; a surveyor should be wary of certifying to the existence or nonexistence of wetlands. This is addressed in item 19 on Table A of the Code by requiring the location of wetlands areas as delineated by appropriate governmental authorities, i.e., federal, state, or local maps showing wetland areas.

Much of the remainder of this clause deals with items that would be shown by virtue of the following clause (iii) anyway. These items often appear in various forms of certification because the attorney who drafts the form had a "horror story" due to a surveyor's failure to show a railroad siding or some other particular improvement. To the extent that these items are appropriate for the surveyor to certify, they are covered in the Code: "on the ground" in the first paragraph of section 5 of the text, "improvements" in paragraphs 5.B, C, and D of the text and item 8 of Table A, and cemeteries in paragraph 5.F of the text.

(iii) this Survey shows the location of all rights-of-way, easements, building lines, setbacks, height restrictions, side-yard requirements, parking areas, and other conditions which appear in a title report by the Title Company dated _________, 20___, or which were apparent from inspection of the Premises or which are imposed by law or ordinance affecting the Premises;

*Note:* These items are covered in paragraph 5.E of the text of the Code. It is not clear what the language "conditions . . . imposed by law or ordinance affecting the Premises" is supposed to cover. It could include compliance with building and electrical codes, or the adequacy of the waste disposal system. One should not ask a surveyor to opine on matters outside his professional expertise, especially if such a request is inadvertent.

(iv) except as shown on this Survey, there are no buildings, other structures, or improvements which are located on the Premises and which encroach upon any rights-of-way, easements, building lines, parking areas, or adjoining property lines.

*Note*: The determination of whether an improvement "encroaches" on any easement or adjoining properties is a legal determination. The surveyor can only show the relationship of the improvements to the record liner. This is covered by the Code, in particular, at paragraph 5.C(ii) and (iii) of the text.

(v) except as shown or noted on this Survey, the existing improvements on the Premises do not violate any setback, building line, side-yard, height, bulk, frontage, density, dimensional, or parking restrictions and/or requirements which affect the Premises;

*Note:* Once the certifications go beyond describing or plotting setback lines and height requirements, etc., the issues become cloudy and frequently should be addressed by a zoning opinion or letters from the zoning authorities. For example, compliance with density and bulk restrictions may involve measurements of interior net rentable or a determination of net usable floor area which the zoning ordinance may not define how to measure. If the survey is of a corner lot, the zoning ordinance may not be clear as to whether the front setback applies to both street frontages or which of the interior lines are considered "rear" or "side" lines. On the issue of height, it often is not clear under the zoning regulations what the "height" of a structure is: the peak of the roof, the top of a chimney or antenna? Height above what: the center line of the street, the ground at the base of the first floor, sea level? When these items are covenants in a recorded document or shown on a filed map, these questions cannot be answered definitively if it is not clear from the recorded document. If such information is required from the surveyor, one should specifically negotiate and delineate such a requirement rather than leaving it to a "catchall" clause in a standard certification. The lawyer should bear in mind that researching these ambiguities is beyond the scope of a surveyor's expertise. Zoning information appropriate for a surveyor to certify is adequately covered by item 6 of Table A of the Code.

(vi) the Premises has adequate ingress and egress to ________via direct access driveways;

*Note*: What is "adequate" ingress and egress? Is it "adequate" if it is suitable for the needs of tenants? For delivery trucks? For fire trucks? If it complies with traffic department requirements? A surveyor should hesitate to certify to this clause; instead, a statement that there is vehicular access to a physically open public street as shown on the survey would be more appropriate. This requirement is covered by paragraph 5.B(i), (ii), and (iv) of the text of the Code for legal access and 5.B(iii) for physical access.

(vii) the Premises has the following U.S. Department of Housing and Urban Development Flood Plan Classification: ________________________________, as reflected by Flood Insurance Rate Map Panel #________, dated ________, which such panel covers the area in which the Premises is situated;

*Note*: The certification should make clear that this determination is made by graphic plotting only. Any more precise location requires a separate "vertical control" survey which, in most cases, is probably not necessary. Item 3 on Table A of the Code adequately addresses flood zones.

(viii) the Premises is subject to the zoning jurisdiction of ______________, and it has the following zoning classification: ____________;

*Note:* This is covered by the Code in item 6(a) of Table A of the Code.

(ix) the present and/or projected use of the Premises does not violate the zoning classification of the Premises;

*Note*: One should not ask the surveyor to opine on the use of the property (and particularly not its projected use, assuming the surveyor even knows what that is). While physical compliance with zoning setback and height requirements is often an appropriate matter for a surveyor to certify (but see the above *Note* to clause (v)), questions of permitted use are for a zoning attorney or architect to answer. Item 6(b) on Table A of the Code is a more appropriate formulation.

(x) the Premises has ___________ parking spaces as shown on this Survey;

*Note*: This is covered by item 9 on Table A of the Code.

(xi) the number of parking spaces required by the applicable zoning local ordinance for such improvements based on their intended use is __________;

*Note:* Like setback lines and use restrictions, unless the required number of parking spaces is based on the area of the entire parcel, it is difficult to certify to the required number of spaces. Parking requirements often depend on the number of tenants, or square footage of retail space, etc., and this is a determination that a land surveyor typically cannot make. At a minimum, it would require measurements of internal spaces and inspections of apartments, all of which are outside the scope of a title or boundary survey. If the client is concerned about adequate parking or intends to change the use after he or she buys the property, a zoning opinion or architect's certification would be more appropriate. Item 9 on Table A of the Code more appropriately addresses parking.

(xii) the Survey shows the direction and flow of streams, rivers, and surface drainage systems and the location of rubbish fills, filled in wells, and seepholes;

*Note*: This language is very broad and could serve to hold the surveyor responsible for locating potential hazardous materials dump sites. Further, many of these items may not appear from a surface inspection. An environmental assessment may be a more appropriate way to ascertain these facts. If this clause remains, the surveyor would be wise to include an express disclaimer of any liability with respect to locating or noting any evidence of hazardous materials. To the extent there are water flows on the property, this is adequately covered by paragraphs 5.B(g) and 7 of the text of the Code.

(xiii) the Survey describes the frame, roof, and exterior finish of all structures;

*Note*: Such descriptions are not typically a part of a surveyor's expertise. If this information is critical, one might more appropriately gather it through an engineering or structural report on the condition of the improvements. Not surprisingly, this is not covered by the Code.

(xiv) the Survey shows the nature and location and all utilities serving the Premises including manholes, catch basins, valves, vaults, or other surface indications of such structures, including abandoned lines, wire and cables (including their function), wire bearings, or guidepoles on or within ten feet of the property, together with pole numbers;

*Note*: This information is not typically necessary in a "title" survey. If the utilities enter the property from a nonpublic road or service adjoining properties, they would appear as physical evidence of easements under clause (iii). If this detail is required for a particular purpose, e.g., construction, the certification should limit this clause to aboveground utilities. With respect to subsurface utilities, the certification should be qualified to reliance on information provided to the surveyor by the appropriate utility company or municipality. Without excavation of the site, the surveyor can only rely on these two sources. Item 11 of Table A of the Code deals with utilities in the title survey context.

(xv) all physical evidence of boundary lines and lines of possession or occupancy have been shown and proper notation made where in conflict with the legal description of record;

(xvi) except as shown on the survey, there are (a) no discrepancies between the boundary lines of the subject property as shown on the survey and as the same are described in the legal description of record and (b) no deficiencies in the quantity of the land described in the legal description of record; Deficiencies in quantity is jargon for shortages in area

(xvii) except as shown on the survey, there are no gaps, gores, or overlaps between the boundaries of the subject property and the boundaries of any adjoining parcels or roads, highways, streets or alleys, and all parcels which comprise the subject property are contiguous; and

*Note*: Although clauses (xv), (xvi), and (xvii) above are perhaps the most important certifications that a surveyor makes (after all, surveyors are recognized as professionals because of their peculiar expertise in boundary line location), most local and state surveying codes of practice (and the ALTA/ACSM Code prior to its 1988 revision) do not require investigation of adjoining property descriptions, reconciliation of potential overlaps or gores, or correction of descriptions that do not close. Many codes do not even require reconciliation of physical possession and record lines and allow the surveyor to adopt possession as the boundary lines. Clauses (xv), (xvi), and (xvii) above are properly insisted upon; however, paragraphs 5.C(i), 6.B(vii), 6.B (viii), and of the text of the Code adequately address these issues.

(xviii) this Survey is an "ALTA/ACSM Land Title Survey" and has been prepared in accordance with the "Minimum Standard Detail Requirements for Land Title Surveys" adopted by ALTA and NSPS in 2011.

*Note*: This reference to an ALTA/ACSM Land Title Survey is incorrect and should not be used or accepted. The 2011 Code requires that a survey is *not* an ALTA/ACSM Land Title Survey unless it contains the following certification, *unaltered* except for changes that are required by federal, state, or local laws or regulations governing title surveys:

"This is to certify that this map or plot and the survey on which it is based were made in accordance with the 2011 Minimum Standard Detail Requirements for ALTA/ACSM Land Title Surveys, jointly established and adopted by the ALTA and NSPS, and includes items _______ and Table A thereof. The fieldwork was completed on ___________."

SURVEYOR ______________________

Date:_________________________ By:_________________________

[SEAL] Title: _______________________

*Note: The law of many states requires that a survey bear the ink or embossed seal of the surveyor who prepared it. In these states, an unsealed survey cannot be relied on as the unaltered work product of the surveyor who prepared it.*

## Conclusion

With appropriate references to the required items on Table A, the 2011 ALTA/ACSM certification alone should be sufficient for any lender (including FNMA) or title company. It may or may not be sufficient for an owner or developer, depending on the purpose of the survey and the owner's intended use of it.

A final caveat: when all is said and done, the attorney and the surveyor must be clear about the purpose of the survey and the goal of the client. Everything in this article (and in the Code) is aimed at title surveys. An ALTA/ACSM survey will not meet the needs of a client planning on constructing a new building.

# 7 Certification from a Surveyor's Point of View

Robert W. Foster, PE, PLS

Surveyors are frequently called upon to provide certifications for their work. To most surveyors, their seal and signature are sufficient to indicate that they stand behind their reports and statements. To an increasing number of clients and agencies, however, the certification is necessary as a specific statement of reliability of the surveyor document to which it refers.

The surveyor is a reporter of facts before he is a supplier of opinions. The word *certify* means to "confirm," "attest to," "verify," "affirm," "substantiate," "guarantee," or "warrant."[1] According to Black's Law Dictionary, a certification is "the formal assertion in writing of some fact."[2] Clearly, a certification in this context must not involve a statement of opinion. When a surveyor measures certain features of a parcel of land and reports them on a plan or plat of the site, that surveyor is a reporter of facts. When a surveyor offers an opinion as to the reliability of the evidence determining the property lines of the parcel, or when the surveyor offers an opinion as to compliance with zoning for instance, the surveyor is no longer merely a reporter of facts. The surveyor's certification should never be in the form of or contain an opinion.

1. Oxford Dictionary and Thesaurus (American ed. 1996).
2. Black's Law Dictionary (abridged 5th ed. 1983).

A few typical forms will demonstrate the dilemma of the surveyor in dealing with the certification statements often requested by clients.

## I. Land Title Insurance Survey Certifications

There has been a succession of standards entitled "Minimum Standard Detail Requirements for Land Title Surveys," beginning with the 1962 standards published by the American Title Association (ATA) and the American Congress on Surveying and Mapping (ACSM). The standards were revised unilaterally by ACSM in 1979 and jointly by ACSM and the American Land Title Association (ALTA) in 1986, 1988, 1992, 1997, 1999, and again in 2011. Unfortunately, the land title industry is not uniformly aware of these revisions, and surveyors are still asked occasionally to work to the 1962 standards or one of the subsequent versions and to execute a certification accordingly. Surveyors are advised to make reference *only* to the most recent version of the standards.

The 1962 ATA/ACSM standard made reference to exactness in surveys. Exactness, perfection, and "error-free results" are words and phrases inappropriate to a description of the surveyor's work. There is no such thing as exact, perfect, or error-free measurement. In 1979, ACSM revised the standard, eliminating the inappropriate language but requiring the surveyor to report "maximum positional tolerance of corners." Unfortunately, the standard did not define the term, nor was the term found in standard surveying texts and books of definitions. Furthermore, the 1979 standard was approved only by ACSM and never became a document published jointly by the two associations.

The revisions to the standards as adopted by both associations over the following thirty-two years included a restatement of measurement standards as "relative positional precision" with a definition of the term. The modern standard includes required features of a site to be included in the survey as well as certain "optional survey responsibilities and specifications," titled Table A. The standard calls for the following statement of certification:

> *This is to certify that this map or plat and the survey on which it is based were made in accordance with the 2011 Minimum Standard Detail Requirements for ALTA/ACSM land Title surveys, jointly published and adopted by ALTA and NSPS, and includes items _____ of Table A thereof. The fieldwork was completed on _____.*

(The National Society of Professional Surveyors (NSPS) is a member organization of ACSM.)

A note to Table A carries the caveat that the surveyor cannot make a certification on the basis of an interpretation or opinion.

The certification statement acknowledges responsibility of the surveyor for the full scope of the standard including other federal, state, or local rules regulating the practice of surveying; the "normal standard of care" of surveyors; boundary resolution; compliance with the measurement standards; consideration of a complete set of "Record Documents" defining the title to the property and any entitlements, servitudes, and covenants; and relevant features of the site relating to monuments, rights-of-way, lines of possession, buildings, easements and servitudes, cemeteries, and water features. It is an extensive, all-inclusive range of responsibilities assumed by the surveyor when he or she signs the required certification.

## II. Alternate "Surveyor's Report"

Appendix A to this chapter is an example of a surveyor's report typical of those requested by some banks and title insurance companies. It is more than a report; the entire form is a certification statement.

Surveyors should reject item 1 of this form. One cannot reasonably expect a surveyor to report the nonexistence of "hidden" features, i.e., features that are not normally visible upon inspection. The surveyor may verify the *existence* of underground drains, and water, sewer, gas, and oil pipes when facilities such as manholes and valves are observable at the surface. There is no effective way, however, for the surveyor to determine that such underground features *do not* exist. In many instances, the same is true of cemeteries and family burial grounds (item 3) and abandoned rights-of-way and old highways (item 1).

Item 9 of this form is vague and rests more within the expertise of the engineer or architect than of the surveyor. A proposed change in a street line (item 10) is information not discoverable by an examination of the property. "Indications of recent street or sidewalk construction or repairs" is vague and indefinite.

Item 12 would require the surveyor to rely upon the work of the title examiner. Rather than respond to the question as asked, the surveyor should list whatever restrictive covenants have been made available to him, cite their source, and then deal only with the question of compliance as regards the location and extent of the improvements. Surveyors should refuse to comment on compliance with restrictive covenants that refer to "use" and "occupancy." Surveyors should address only positional or dimensional requirements of restrictive covenants.

## III. The Generic "Surveyor's Certificate"

Appendix B to this chapter is a copy of a certification written for a specific project but is similar in form and substance to forms that surveyors often encounter. The statement uses the phrases "true and accurate" and "true and correct." In the context of surveying, these phrases mean acceptable accuracy. To everyone else, however, the phrase "true and correct" means exact. Because this certification conveys information to non-surveyors, it is incumbent on the professional to avoid terms and phrases that convey the wrong message. The reference should be *to appropriate precision and accuracy.*

The surveyor should qualify the statement on encroachments, protrusions, and improvements on the property so as to specify only observable encroachments, protrusions, and improvements. Further, because a surveyor cannot know that the title examiner has discovered "all recorded easements," the surveyor should certify only that the plan shows specific easements as shown thereon. Finally, the last paragraph of this certification might be construed as an express warranty. The prudent surveyor should not state that the purpose of the certification is to induce the lender to extend credit and to induce the insurer to issue a policy of title insurance. Surveyors should limit their activities to the reporting of facts and leave persuasion to others.

## IV. Conclusion

Lawyers and professional liability insurers who deal with surveyor liability issues advise caution when executing certifications containing the problems discussed here. They also advise against certifications that claim a standard higher than the normal standard of care of surveyors, certifications that are inconsistent with contractual duties, certifications that create guarantees or warrantees, or certifications that create unacceptable or uninsurable exposure to risk. Surveyors have learned to be as conservative as the experts who advise them and the companies that insure them. Above all, surveyors must be aware of the broad extent of responsibilities they assume when making certification to completion of duties that are all-inclusive and may be beyond the normal scope of their work.

# APPENDIX A
## Surveyor's Report
## (to be filled out and signed with all surveys)

THIS IS TO CERTIFY that on ________, 20____, I made an accurate survey of the premises standing in the name of __________, updated at [City], [County], [Street] briefly described as: _________________________and shown on the accompanying survey entitled: ____________________

____________________________________________________________

I made a careful inspection of said premises and of the buildings located thereon at the time of making such survey, and again on ______, 20___, and at the time of such latter inspection I found ______________ to be in possession of said premises as [tenant or owner].

I further certify that as to the existence or non-existence of the following at the time of my last inspection:

1. Rights-of-way, old highways, or abandoned roads, lanes, or driveways, drains, sewer, water, gas, or oil pipe lines across said premises: __________.
2. Springs, streams, rivers, ponds, or lakes located, bordering on, or running through said premises: ___________________________________________

   _____________________________________________________
3. Cemeteries or family burying grounds located on said premises (show location on plat): ____________________________________________

   _____________________________________________________
4. Telephone, telegraph, or electric power poles, or wires or lines overhanging or crossing said premises and serving other property or properties: ____
5. Joint driveways or walkways; party walls or rights of support; porches, steps, or roofs used in common or joint garages: _____________________
6. Encroachments or overhanging projections (if the buildings, projections, or cornices thereof, or signs affixed thereto, fences or other indications of occupancy encroach upon or overhang adjoining properties, or the like encroach upon or overhang surveyed premises, specify all such):

   _____________________________________________________
7. Physical evidence of boundary lines on all sides (be specific): _________
8. Is the property improved? __________________________________________

   (a) Building is: brick ( ); clapboard ( ); other ( ):______________

   Building is: one-story ( ); two-story ( ); other ( ): _____________
9. Indications of building construction, alterations, or repairs within recent months: _______________________________________________

   _____________________________________________________
10. If new improvements under construction, how far have they progressed?
11. Changes in street lines either completed or officially proposed: Are there indications or recent street or sidewalk construction or repairs?

12. Are the abutting streets or roads maintained by public authorities? _____
13. If the surveyed premises are subject to restrictive covenants, do the improvements, use, and occupancy comply with such? (If the premises are subject to restrictive covenants, has the examining attorney furnished you with a verbatim copy of them?)______________________________

______________________________________
Civil Engineer or Surveyor

NOTE: In all cases where there are encroachments, support easements, party walls, etc., they should also be denoted upon the map of your survey.

# APPENDIX B
## Surveyor's Certificate

The undersigned Registered Surveyor (the "Surveyor") hereby certifies that (a) this plat of survey and the property description set forth hereon are true and correct and prepared from an actual on-the-ground survey of the real property (the "Property") shown hereon; (b) such survey was conducted by the Surveyor, or under his supervision; (c) all monuments shown hereon actually exist, and the location, size, and type of material thereof are correctly shown; (d) except as shown hereon, there are no encroachments on the Property or protrusions there from, there are no improvements on the Property, there are no visible easements or rights-of-way on the Property, and there are no visible discrepancies, conflicts, or shortages in area or boundary line conflicts; (e) the size, location, and type of improvements, if any, are as shown hereon, and all are located within the boundaries of the Property and set back from the Property lines the distances indicated; (f) the distance from the nearest intersection street or road is as shown; (g) the Property has access to and from a public roadway; (h) all recorded easements have been labeled and correctly platted hereon; (i) the boundaries, dimensions, and other details shown hereon are true and correct; and (j) the Property is not located in a 100-Year Flood Plain or in an identified "flood prone area," as defined by the U.S. Department of Housing and Urban Development, pursuant to the Flood Disaster Protection Act of 1973, as amended.

The Surveyor expressly understands and agrees that (a) this Certificate is made to induce ______________ ("Lender") to extend credit secured by a mortgage lien covering the Property and to induce ____________________________________ ________________ ("Title Company") to issue a policy of title insurance insuring the validity and priority of such mortgage lien; (b) both Lender and Title Company are entitled to rely upon this survey as being true and accurate in all respects and upon this Certificate being true and accurate; and (c) the consideration paid to the Surveyor for the preparation and certification of such survey has been paid, in part, for the benefit of Lender and Title Company and in anticipation of their reliance thereon.

______________________________

[SURVEYOR'S SIGNATURE]

Date: _________ By: _____________

[SEAL] Title: ___________

# A Code for Title Insurers[1]

# 8

Mitchell G. Williams

Codes of practice for the land surveying profession are common. Most state surveyor's associations have promulgated surveying standards; some states have enacted practice standards as state law or adopted them as regulations through their licensing and registration boards. There are also national sets of standards, the most well known and widely used being the Minimum Standard Detail Requirements for ALTA/ACSM Land Title Surveys, which has been in existence since 1962 in one form or another. Except in those states with statutory standards, few surveyors admit on the face of their survey maps that the survey was made in compliance with any particular code or any minimum standards at all. In some areas of the country, there is considerable resistance to, and occasionally outright refusal of, a request to certify a survey to some specified code of practice or to provide any kind of certification at all.

Such a response is not only puzzling, but is also certain to raise doubts in the minds of clients and their attorneys about the scope and quality of the survey rendered. For the most part, laymen and lawyers do not know what a surveyor does (or should do) in order to deliver the finished map of survey. Nor should they be required to find out. The land surveyor is, in most states, a licensed professional, and clients and attorneys should be able to rely on the surveyor's exercise of his or her professional judgment. Although a surveyor may feel it is only necessary to "certify" his work, secure in the knowledge of what constitutes the "standard of care" of his fellow professionals, his clients do not share this security of knowledge. As a result, attorneys and their clients should insist on their surveys

1. An earlier version appeared in Legal Topics in Boundary Surveying: A Compendium (Wendy Lathrop ed., 1990).

being done in accordance with a particular code, and that the surveyor certify his or her work in accordance with an appropriate set of standards that they can deliver to their client together with the finished map.

The use of an appropriate code of practice will inform the client of the scope of the surveyor's work and what the client can expect to find and *not find* in the survey. An owner of property should not rely on a survey to indicate the presence or location of interior cemeteries if the survey clearly indicates that it was done in accordance with the ALTA/ACSM standards. Conversely, the selection of a code by the *client* clearly indicates to the surveyor the scope of work that the client expects. A request for an ALTA survey under the current code should discourage the surveyor from furnishing a "mortgage map" drawn from the front seat of his car.

Specifying a set of standards also defines the standard of care to which the work should be done. This can protect the surveyor who was hired to do a pure topographical survey from a client who later builds a twenty-story building in reliance on lot lines that were shown only for reference purposes. Similarly, this can protect the client who requested a boundary survey and assurance of contiguity from the surveyor's later claim that the "custom" in the area was not to research the descriptions of adjoining properties.

Codes of practice also serve as evidence of the level of care that the profession itself considers to be the minimum performance expected of a surveyor. They provide attorneys with the opportunity to look at the surveying profession's own requirements for the scope and quality of a survey. Even those surveyors who refuse to comply with the code common in their state or area are likely to be judged by its standards. Of course, if the code itself is a poor one—and many of them are—there is a risk that the standard of care will be too low. In fact, this has happened in some areas where "mortgage maps" and "tape surveys" have become the standard for title surveys to the debasement of the profession as a whole.

There are two major classifications of practice codes, one based on the promulgating authority and the other on subject matter. Codes can be statutory, such as Ohio's standards for boundary surveys (Standard for Boundary Surveys in the State of Ohio, OHIO ADMIN. CODE ch. 4733–37); regulatory, such as Connecticut's code of practice (Minimum Standards of Accuracy, Content and Certification for Surveys and Maps, Rules and Regulations of State Board of Registration for Professional Engineers and Land Surveyors, § 20–300b-1 through 6–20); or established by private associations, such as the ALTA/ACSM standards. With respect to subject matter, codes can be generic and apply to all surveys, such as Florida's regulatory requirements (Board of Professional Land Surveyors and Mappers, Minimum Technical Standards, Department of Business and Professional Regulation, ch. 61617–6); specific to one type of survey, such as Ohio's standard for boundary surveys; or "comprehensive," in which a single code addresses the particular nuances of each type of survey, such as the Indiana code of practice for surveys (Land Surveying; Competent Practice, 865 IND. ADMIN. CODE 12).

From another perspective, the existing codes of practice exemplify two distinct but not inconsistent approaches. One could be called the "technical" approach. A good example is that of the Connecticut Code. Most of it deals with the minutiae of tape standards, the number of repetitions of angles, closure of field traverses, and so forth. The other is a more "conceptual" approach, with the emphasis on "what to measure" rather than "how to measure." Florida's statutory directives require the surveyor to determine the position of the boundary lines based on the record lines and to indicate all overlaps and erroneously located lines. There is no pure example of either a "technical" or "conceptual" code, but these approaches are found in very different proportions. For example, Connecticut's is primarily technical, Indiana's title survey code is primarily conceptual, and the 2011 ALTA/ACSM Code strikes a good balance between the two approaches. None of these approaches is good or bad in and of itself; however, in light of the purposes for which codes of practice exist, a "conceptual" code is preferable. The "technical" code is of minimal use to the end user, whether it be an attorney, a title insurer, a developer, or a property owner. The client does not know what an error ellipse is, let alone how to interpret one, but the client does know that he wants the surveyor's professional opinion of the location of his property lines. In addition, surveying is a growing and constantly changing profession. No one is served by being locked into an outdated technology of measurement by a "technical code." Technical standards requiring specific measurement techniques frequently lag behind advances in surveying technology and discourage or may prevent the use of electronic distance-measuring equipment and the use of the global positioning system.

The ultimate user of the survey has no way to judge the appropriateness of any of these techniques. In the majority of states, surveyors are licensed professionals who must have extensive theoretical and practical experience and knowledge of the methods of measurement and determination of boundary lines. Surveyors are recognized as expert witnesses with respect to property lines and are regarded as having specific expertise separate from and beyond that of the civil engineer. The status of a licensed professional should assure competence in the technique of measurement. To be of greatest use, a code of practice should be a statement of what is to be done by the surveyor. It should inform the client of what he can expect from the surveyor, and the surveyor of what the client expects from him.

The 2011 ALTA/ACSM Code solves the problem with the "technical" approach by defining the precision required in the location of the boundary lines and corners, leaving the method of obtaining of that precision to the surveyor's professional judgment. Because it is a title survey code, however, it cannot serve as a model for the profession as a whole without modifications for an owner or developer.

Except for the statutory standards, most codes of practice were initially promulgated by the title insurance industry, either nationally or by local state title insurance associations, and then adopted by various surveying societies. The title insurance industry is a significant end user of the surveyor's expertise and has a particular interest in getting surveys on which title insurers can rely. The title insurer's interest is a limited one, however, and a title survey that is adequate from an insurer's point of view is not necessarily enough for an owner or developer. Unfortunately, some of the surveying profession's attempts to codify its practice appear to have been little more than reactions to a fear that the title industry would usurp the role of surveyors' associations. ACSM had promulgated "Technical Standards for Property Surveys" in 1946, yet in 1962 it adopted as its own the standards for title surveys of the then American Title Association. Rather than being adapted to meet the needs of the title insurance industry, the 1946 Technical Standards became a footnote in the ALTA code: they were "recommended." Even this lip service to the ACSM property survey standards had been eliminated from the ALTA/ACSM Code by 1997. As a result, the only widely recognized surveying code on a nationwide level is a title survey code, and even that is not a product generated by ACSM, but a joint enterprise with ALTA. Some state associations simply adopted the local title association code for title surveys with minor amendments and called it a "Code of Practice for Land Surveys." To the unsuspecting, it is a generic code; in reality, it is for title insurance purposes only.

Title insurance is a risk-averse industry. A title policy excepts from coverage, among other things, all title defects that the company has discovered of record, all defects of which the insured may have had knowledge and failed to disclose to the insurer or which result from the acts of the insured, and in most policies (particularly owners' policies), "any state of facts an accurate survey may show." The sole purpose of a title survey is to allow the title insurer to remove this so-called "survey exception" from the policy.[2] The survey exception is usually not removed, however, but replaced with a "survey reading" that *excepts* from insurance coverage the specific defects that the survey reveals. If the survey shows that the lot does not close, that there are overlaps or gores with adjoining properties, or that there are encroachments of improvements or easements, these defects are simply not insured against. In addition, there are many things that an accurate survey would show that the title policy does not and cannot insure against, because they are not deemed to be title defects (e.g., zoning compliance, setback requirements, and utilities).

The decision to eliminate or limit the survey exception is both a legal and economic one. On a lender's policy, the exception is very likely to be taken out

2. As an aside, the 2006 ALTA form of owner's policy includes in section 2(c) of "Covered Risks," insurance coverage with respect to "any encroachment, encumbrance, violation, variation, or adverse circumstance affecting the Title that would be disclosed by an accurate and complete land survey of the Land. The term encroachment includes encroachments of existing improvements located on the Land onto adjoining land, and encroachments onto the Land of existing improvements located on adjoining land." This coverage is negated or limited by the existence of a survey exception or survey reading, respectively, in Schedule B of the policy. The new insuring language does not increase the coverage provided under the prior ALTA policy forms, but was added in response to a number of court decisions under the prior policy forms that held that even without any survey exception on Schedule B, the policy did not provide affirmative coverage.

because most sophisticated mortgage lenders and their attorneys insist that it be taken out. On owner's policies, it often costs more in premiums to have the survey exception removed, and many attorneys do not realize its significance. In the event of a loss due to a boundary line mislocation or encroachment, the title company's liability can be limited to the face amount of the policy, which may be far below the market value of the property "lost." The owner faces the loss of possession, costs to remove encroachments, etc., which may far exceed the insurance coverage. Not only are the title insurer's survey concerns different from a property owner's, but its risk of loss can also be substantially lower.

This can be seen in the wide range of acceptable ways to remove the survey exception: visual "inspections" by title company employees, the notorious "mortgage map," the "location service" provided by Chicago Title (until the execution of a consent decree in April 1986 with the Professional Land Surveyors of Ohio), or affidavits from the current owners that they have not had a boundary dispute with any of their neighbors within the statutory period for adverse possession. Each of these will remove or limit the survey inspection, but none of them could be relied upon in order to erect a fence along a property line. It is from this perspective that the ALTA/ACSM standards are lacking for the owner/developer and fail to serve the legal profession as a model standard of care by which all surveys should be judged.

One of the problems with a title survey code as a "model" code is that it is both over- and under-inclusive. Suppose that a surveyor is hired specifically to locate and stake a single property line for a fence. If he certifies that his work is in accordance with the ALTA/ACSM Code, will he be liable if he does not note the existence of a cemetery on the property that he noticed while running the line? In this case, the code imposes additional burdens and costs that are unnecessary for the project. In contrast, if the legal description of the property does not mathematically close, the surveyor can still stake the line and certify compliance with the ALTA/ACSM Code (which only requires that the failure of the legal description to mathematically close be noted on the survey map), even if it is later determined that the line staked contained the error which prevented closure. The risk of liability for the surveyor is increased at the same time that the ability of the client to rely on the survey is decreased.

Title surveys, construction surveys, boundary surveys, building permit surveys, taking maps, subdivision surveys, and all other types of surveys in which the locations of improvements or new property lines are to be related to existing boundary lines require, as a fundamental datum, the location on the ground of those boundary lines. An owner expects, at a minimum, that the survey map he has ordered will indicate, to the best of the surveyor's ability, the location, dimensions, and extent of his property. This requires the surveyor to research and analyze the deed for the client's property, the record description of all adjoining properties (even in the so-called "metes-and-bounds" states, no property stands alone, but only exists in relation to the surrounding parcels), all available maps, whether for subdivisions or highway rights-of-way, whether or not they are recorded in the land records, and certificates of title. It requires the physical location of all evidence in the field, whether monuments, stakes, physical improvements, or lines of possession or occupation. The record data must be reconciled both with itself and with the physical evidence. Overlaps and gores must be identified, errors of closure discovered, misalignments of record and occupied lines delineated. Out of this mass of often confusing and inconsistent data, the surveyor must establish, in his professional judgment (and not as an advocate of his client), the most probable location of the boundary lines of the property in question.

Due to both the nature of a title company's role as an abstracter and its specific requirements for a "title" survey, an ALTA/ACSM survey need not involve all of the above determinations. The surveyor is under no duty to search the record or even obtain a description of the client's property: he is only bound to work from the information provided by the client. In most cases, an owner has neither the expertise nor the ability to ferret out all the record information needed. This is usually one of the reasons the owner has ordered a survey in the first place. The ALTA/ACSM Codes prior to the 2011 revision did not require the surveyor to determine the boundary lines or reconcile overlaps, gores, discrepancies between occupied and record lines, or errors of closure; it was sufficient simply to indicate their existence. Even if the description of the client's property failed to close, the Code prior to 2011 did not require that the surveyor indicate the most likely source of the error and provide a corrected description; simply noting the failure to close was sufficient.

The range of potential uses for surveys is immense, and each has specific requirements beyond the needs of a title insurer. The zoning classification of the property, including use and setback, height and bulk restrictions, the location of all existing utilities, the staking of the actual corners of the property or of improvements to be built, topographical information and elevations, parking areas, location in a federally designated flood hazard zone, state plane coordinates for property corners, and other items may be relevant for the needs of a particular client. The cachet of an ALTA survey should not blind the owner, his attorney, or the surveyor to the need for a careful consideration of the purpose of the survey and the requirements that must be met. The list of additional items attached to the ALTA/ACSM Code is an attempt to expand the utility of the Code to cover surveys for other than title insurance purposes, but it is ultimately only a stopgap.

The 2011 ALTA/ACSM Code is the first major revision of the code since it was first adopted by ALTA and ACSM in 1962. It addresses a number of the problems with the prior versions of the Code, but also states that its purpose is to allow a title insurer to insure title clear of survey matters not shown on the survey. But even if there were a "universal code" that covered all types of surveys, it would still be necessary for attorneys to be well informed about the scope of the survey work needed by his client and the provisions of any code of practice to which a survey is done. Most important, there should be clear communication with the *surveyor* about the needs of the client. The survey should not be treated as an afterthought or an adjunct of a title policy. Lawyers and clients cannot rely on the title insurance industry, with its specific needs and cost-benefit philosophy, to define the scope and quality of surveys.

# Surveys and Title Insurance

# 9

David E. Snediker[1]

## I. Title Insurance

Title insurance policies are contracts of indemnity that, subject to the policy terms and conditions, "protect" policyholders against actual monetary losses or damages up to the limit of the policy coverage. Title insurance policies are not guarantees that automatically obligate the insurer to pay money if a title defect is later discovered. The insured must actually sustain a loss or damages based on a title defect for which coverage is provided, and the insurer has the option to either pay money to the insured or litigate against or settle with third parties to protect the insured's interest.[2]

### *A. Policy Forms*

#### 1. ALTA Policy Forms

The American Land Title Association (ALTA) adopted new policy forms in 1992 to create greater uniformity in coverage nationally. The 1992 ALTA forms (the 1992 Forms) were revised again in 2006 (the 2006 Forms). The 2006 Forms (like the 1992 Forms) are subject to differing regional exceptions for water rights, mechanic's liens, rights of parties in possession, and other "off-record" issues. Title

---

1. David E. Snediker is a partner in the real estate department of the New York office of Paul Hastings LLP. Lisa A. Chaney and Kimberly A. Gordon, associates at Paul Hastings, also contributed to this article.

2. Conditions 4 and 5 of the 2006 ALTA Forms, http://www.titlelaw-newyork.com/ALTA/2006OwnersPolicy.pdf, http://www.titlelaw-newyork.com/ALTA/2006LoanPolicy.pdf (last visited Nov. 4, 2011).

companieswill delete or "endorse over" many of the ALTA regional exceptions if the insured provides appropriate documentation (and, in some cases, additional premiums). Although there are some modifications in a few states, the 2006 Forms are now available in at least forty-five jurisdictions.[3] Texas, however, is a notable exception because it requires title insurance companies to use policy forms adopted by the Texas State Board of Insurance.[4] In California and several northwestern states, both ALTA and California Land Title Association (CLTA) policies are available.

## B. Coverage

### 1. The 1992 Policy Forms

Subject to stated exclusions from and exceptions to coverage as well as the policy's general conditions and stipulations, the basic 1992 ALTA lender's policy insured against loss or damage by reason of:

> (1) title to the estate or interest described in Schedule A being vested other than as stated therein; (2) any defect in or lien or encumbrance on the title; (3) unmarketability of the title; (4) lack of a right of access to and from the land; (5) the invalidity or unenforceability of the lien of the insured mortgage upon the title; (6) the priority of any lien or encumbrance over the lien of the insured mortgage; (7) lack of priority of the lien of the insured mortgage over statutory liens for services, labor or material (a) arising from an improvement or work related to the land which is contracted for or commenced prior to the Date of Policy; or (b) arising from an improvement or work related to the land which is contracted for or commenced subsequent to the Date of Policy and which is financed in whole or in part by proceeds of the indebtedness secured by the insured mortgage which at the Date of the Policy the insured has advanced or is obligated to advance; and (8) the invalidity or unenforceability of any assignment of the insured mortgage shown in Schedule A, or the failure of the assignment shown in Schedule A to vest title to the insured mortgage in the named assignee free and clear of liens.[5]

3. James L. Gosdin, Title Insurance: A Comprehensive Overview, Third Edition 4 (Candace M. Cunningham ed., ABA Publishing 2007).

4. *See* David Tomek & Florence Berkley, *The New Texas Title Insurance Policies,* 30 Tex. St. B. Real Prob. & Tr. L. Rep. 2 (1992) (discussing the Texas title insurance policies approved by the Texas State Board of Insurance on August 16, 1990); *see* also Gosdin, *supra* note 3, at 583–97 (comparing the Texas Owner's Policy and the 1992 ALTA Owner's Policy).

5. 1992 ALTA Forms, http://www.alta.org (last visited Nov. 2, 2011).

## 2. The 2006 Policy Forms.[6]

The basic structure of the 2006 Forms remains consistent with that of the 1992 Forms by providing insurance coverage that is subject to exclusions from coverage, exceptions from coverage (i.e., property specific matters listed in Schedule B), and "Conditions" (which were titled "Conditions and Stipulations" in the 1992 Forms).[7]

The section in the 2006 Forms that describes loss events insured against now refers to them as "Covered Risks" (Covered Risks) and provides more clarity of coverage by expressly identifying in this section certain loss events that were only implied in the 1992 Forms. The 1992 Forms implied that coverage was based on general insuring provisions and carve-outs from loss events excluded from coverage in the "Exclusion from Coverage" section of the policy. The use of carve-outs to exclusions to imply coverage (i.e., reliance on "double negatives") led to ambiguity in interpreting coverage. Indeed, some courts held that the analysis of coverage ends with the insuring clauses, meaning that the exceptions to exclusions from coverage were not recognized as providing coverage.[8]

Covered Risk 2(a), which provides insurance against "defects" in title, illustrates the movement from implicit to express coverage in the 2006 Forms. Though both the 1992 Forms and 2006 Forms insure against "any defect in or lien or encumbrance on title," the 2006 Forms go on to expressly list numerous kinds of defects that could affect title. These defects include, but are not limited to, those caused by forgery, fraud, undue influence, incapacity, lack of authority, improper execution or delivery, and improper recording of documents. Defects also include failure to perform any of the required acts by electronic means authorized by law.[9]

A potentially significant addition to the 2006 Forms is coverage, contained in Covered Risk 2(c), which expressly insures against loss due to any "encroachment, encumbrance, violation, variation, or adverse circumstance affecting the Title that would be disclosed by an accurate and complete land survey of the Land" without the need for a special survey endorsement.[10] Presumably, title companies will, at least for commercial transactions, require surveys to provide this coverage without a special survey exception, and they may also require that a special survey exception be included by having the survey expressly referenced

---

6. *See* 2006 ALTA Forms, http://www.titlelaw-newyork.com/ALTA/2006OwnersPolicy.pdf, http://www.titlelaw-newyork.com/ALTA/2006LoanPolicy.pdf (last visited Nov. 4, 2011). Jeffrey G. Gurren, *2006 ALTA Title Insurance Policies. The Significant Changes,* 24 PRAC. REAL EST. LAW 17, 19–20 (2008).

7. Elliot L. Hurwitz & Randall Kadlec, *ALTA Adopts New Policy Forms for Both Owners and Lenders,* N.Y. L.J., Aug. 21, 2006.

8. Paul L. Hammann, *2006 ALTA Policy and Endorsement Forms* (2006), http://www.firstam.com/ekcms/uploadedFiles/firstam_com/References/Reference_Articles/2006-alta-policy-forms.pdf; *see also* Elysian Inv. Group, L.L.C. v. Stewart Title Guar. Co., 105 Cal. App. 4th 315, 324 (2002 ("It is well established in California that an exclusion cannot act as an additional grant or extension of coverage.") (citing Stanford Ranch, Inc. v. Maryland Cas. Co. 89 F.3d 618, 626–27 (1996)).

9. Gurren, *supra* note 6, at 18–19.

10. GOSDIN, *supra* note 3, at 11–12.

in Schedule B to eliminate coverage for defects shown by a specified survey or, if specific defects shown on the survey are "read in" to the policy, only those defects so noted. This new survey coverage is discussed in further detail below.

The 2006 Forms clarify creditors' rights coverage by adding new Covered Risks 9 and 13, which affirmatively insure against defects in the chain of title due to fraudulent or preferential transfers arising before the insured transaction.[11] This is a key change from the 1992 Forms in which the creditors' rights exclusion applied solely to the transaction covered by the policy.[12] Notably, the 2006 Forms expressly provide coverage for a claim resulting from a court order that either avoids or provides an alternative remedy for any fraudulent or preferential transfer occurring *before* the insured transaction. In addition, if the insured transaction is determined to constitute a preferential transfer due to the related filing being untimely or providing insufficient notice to a purchaser for value or to a judgment or lien creditor, Covered Risk 13(b) in the 2006 Forms provides affirmative coverage to the insured.[13]

The 2006 Forms also provide what is known as "gap" coverage, whereby the title company assumes the risk of any liens or encumbrances that appear between the last effective date of the title certification and the recordation of the insured documents.[14] This coverage, therefore, eliminates the need for a gap endorsement, but title companies can be expected to require the same types of affidavits and indemnities that they previously required for the issuance of a gap endorsement.[15]

## *C. Exclusions*

The 2006 Forms continue to include a section titled "Exclusions of Coverage," but the exclusions in the 2006 Forms no longer contain carve-outs. Notably, the previous carve-outs for police power, eminent domain, and creditors' rights have been moved to the "Covered Risks" section of the 2006 Forms. In addition, the 2006 Forms entirely delete the mechanic's lien exclusion, as Covered Risk 11(a) addresses the issue. Other exclusions, such as "acts of Insured," were not modified in any manner by the new policy.[16]

---

11. Covered Risks 9 and 13 in the 2006 Forms, http://www.titlelaw-newyork.com/ALTA/2006OwnersPolicy.pdf, http://www.titlelaw-newyork.com/ALTA/2006LoanPolicy.pdf (last visited Nov. 4, 2011).

12. Gurren, *supra* note 6, at 19–20.

13. *Id.*; *see also* Covered Risk 13(b) in the 2006 Forms, http://www.titlelaw-newyork.com/ALTA/2006OwnersPolicy.pdf, http://www.titlelaw-newyork.com/ALTA/2006LoanPolicy.pdf (last visited Nov. 4, 2011).

14. Hammann, *supra* note 8.

15. Gurren, *supra* note 6, at 20–21.

16. Gosdin, *supra* note 3, at 21.

## II. Surveys

A land survey is an engineering pursuit in which the surveyor measures and plots dimensions and areas on or near the earth's surface for a stated purpose and in conformity with an agreed-upon standard of accuracy. Surveys help to identify the property involved and uncover relevant information obtained from a physical inspection, which can impact marketability of title to, or the use and enjoyment of, the property. Although surveys are helpful to—indeed critical for—many real estate transactions, surveys are not instruments of conveyance. They provide information regarding physical attributes of the property; they do not create or convey title.[17]

### A. ALTA/ACSM Land Title Survey

Various types of surveys can be performed depending upon the intentions of the parties (e.g., aerial, perimeter, tape location, "as-built," topographical, etc.).[18] The surveyor usually portrays his field efforts graphically in the form of a map, plat, chart, or other agreed-upon medium reflecting the necessary details.[19] An ALTA/ASCM Land Title Survey (commonly referred to as an ALTA survey) is a land survey conducted in accordance with minimum standard detail requirements (the Minimum Standards) jointly adopted by the National Society of Professional Surveyors (a member organization of the American Congress on Surveying and Mapping) and ALTA.[20]

If the surveyor performs an ALTA/ASCM Land Title Survey, it must be prepared in accordance with the then-current "Accuracy Standards" (as defined in the Minimum Standards). The surveyor must also provide at least two sets of prints of a plat or map representing fieldwork conducted on the surveyed ground with the notations and information required by the Minimum Standards. The surveyor must further certify the prints in accordance with the Minimum Standards and affix the surveyor's seal.

### B. Minimum Standards

The Minimum Standards require, among other things, that the person ordering the survey advise the surveyor if an ALTA/ASCM Land Title Survey is desired and

17. *See* Bernard M. Rifkin, *The Land Survey, a Survey of the Law of New York, California, Florida, Illinois and Texas as of 1987, in* TITLE INSURANCE 1989: NEGOTIATING ADDITIONAL COVERAGES 285, 287 (PLI Real Est. L. and Practice, Course Handbook Ser. No. 331, 1989) (discussing the impact of matters discernable from surveys on marketability of title).

18. *See* George Lefcoe, REAL ESTATE TRANSACTIONS 348–50 (4th ed. 2003) (discussing the different types of surveys in greater detail).

19. *See* Gary York, *Surveys in Real Property Transactions,* 5 CAL. REAL PROP. L. REP. 7, 8 (1982) (discussing surveys and survey language in parts 1 and 2).

20. The Minimum Standards were revised, effective February 23, 2011. See http://www.acsm.net for the full text of the Minimum Standards and for changes from the 2005 Minimum Standards (last visited Nov. 2, 2011).

which of the optional items listed in Table A, if any, are to be incorporated. In addition, the surveyor must be provided with the record description of the property to be surveyed and the "Record Documents" (as defined in the Minimum Standards), along with other documentation and information as more fully described in the Minimum Standards.[21]

If an ALTA/ASCM Land Title Survey is requested, the surveyor must perform detailed fieldwork as described in section 6 of the Minimum Standards and deliver a plat or map presented in a prescribed manner that shows evidence gathered during the fieldwork with detailed requirements regarding the depiction of Boundary, Descriptions, Dimensions and Closures and Easements, Servitudes, Rights-of-Way, Access and Record Documents, and a certification in the required form.[22] Table A of the Minimum Standards lists additional items to be included at the request of the client ordering the survey, subject to negotiations between the surveyor and the client, with certain items that were optional under the 2005 Standards now being required by the 2011 Minimum Standards (e.g., "[v]isible evidence of physical access . . . to any abutting streets, highways or other public ways" and "evidence of access to and from waters adjoining the surveyed property. . . .").[23]

Local or state standards may compel variations on the Minimum Standards. When there is a conflict between the local requirements and the Minimum Standards, the more restrictive standard applies.[24]

The Minimum Standards were established to promote a distinct and clear understanding of survey requirements among the insured, the title insurer, and the surveyor. In *Commonwealth Land Title Insurance Co. v. Close, Jensen & Miller P.C.*,[25] the court discussed some of the requirements imposed by the 2005 Minimum Standards. In this case, the title insurer paid claims by the insured owners for loss or damage sustained as a result of encroachments and then sued the surveyor for negligence where the surveyor incorrectly certified that there were no encroachments. In fact, encroachments did exist, but they were not visible because the encroachments were cement floor slabs covered by exterior walls of improvements that were along portions of the boundary lines. While referencing the surveyor's duty, the court concluded, "If a surveyor cannot see a boundary line, he must note the lack of visibility on the survey and he certainly cannot verify that there are no encroachments under such circumstances."[26] Since the surveyor had expressly represented that no encroachments existed without qualifying the representation by noting that only visible boundary line conditions were

21. *Id.*

22. *Id.*

23. Minimum Standards, sections 5.B.iii and vii.

24. *See* Gosdin, *supra* note 3, at 141–43 (discussing the 2005 Minimum Standards in greater depth).

25. Commonwealth Land Title Ins. Co. v. Close, Jensen & Miller, P.C., No. CV065003046, 2008 WL 4926953 (Conn. Super. Ct. Nov. 5, 2008).

26. *Id.*, 2008 WL 4926953, at *4.

being addressed, the court found that failure to be "a breach of the professional standards applied to surveyors" and ordered the surveyor to reimburse the title company for the amount paid to the insured by reason of the surveyor's failure "to exercise that degree of skill and learning commonly applied by the average prudent, reputable member of the profession."[27]

## III. Survey Coverage in Title Insurance Policies

Not every title problem creates liability for the insurer. For example, although certain encroachments (i.e., stoops, walls, windows, etc.) might constitute "title defects," such defects may be permitted by statute, adverse possession, or prescriptive rights. Therefore, if no "actual loss" exists, the insurer sustains no liability.

Prior to the adoption of the 2006 Forms, the insured party was required to provide proof of loss within ninety days of ascertaining the facts giving rise to the loss or damage. If the title company was prejudiced by the insured failing to provide such proof of loss, its obligation under the policy terminated (including any liability to defend the matter requiring such proof of loss).[28] The 2006 Forms shift the responsibility of calculating the loss or damage to the title company. However, if the insurer cannot determine the amount of loss, it may require that the insured furnish a signed proof of loss as a condition of payment. In addition, under the Conditions section of the 2006 Forms, a title company may settle matters with third parties once a claim is raised, and it can pay the policy limit to the insured and cancel coverage.[29]

### *A. Problem Areas under Pre-2006 Forms*

Express coverage for survey matters did not exist under the pre-2006 Forms, with the exception of an affirmation that the property was clear of encumbrances and a statement regarding marketability of title.[30] Consequently, parties frequently litigated whether the insurer could deny coverage based on an exception in the policy relating to survey matters.[31] The standard survey exception existed to refuse from coverage problems that would have been revealed by an accurate survey.[32] The title problems potentially discernable from a "correct" or "accurate" survey include, but are not limited to, the following:

27. *Id.* at *5.

28. Conditions and Stipulations 5 in the 1992 Forms, http://www.alta.org (last visited Nov. 2, 2011).

29. Condition 20 in 2006 Forms, http://www.titlelaw-newyork.com/ALTA/2006OwnersPolicy.pdf, http://www.titlelaw-newyork.com/ALTA/2006LoanPolicy.pdf (last visited Nov. 4, 2011).

30. Gurren, *supra* note 6, at 19.

31. D. Barlow Burke, Jr., Law of Title Insurance § 13 (Supp. 1991).

32. *See* 1440 Empire Boulevard Dev. Corp v. Lawyers Title Ins. Corp., 855 N.Y.S.2d 825 (N.Y. Sup. Ct. 2007) (holding that, even though the insured did not procure a survey, the survey exception did not preclude coverage because the title dispute in question would not be apparent from an accurate

- Discrepancies between recorded and measured courses and distances;
- Discrepancies between measured and "actual" courses and distances (including record and "off-record" information such as the location of improvements, natural monuments, etc.);
- Discrepancies between recorded and measured area;
- Discrepancies between measured and "actual" area;
- Encroachments of buildings, stoops, vaults, walls, fences, landscaping, or other improvements onto property insured or by property insured;
- Encroachments of driveways, streets, highways, and other rights-of-way or easements onto property insured or by property insured;
- Appurtenances such as access, water, and utility rights; and
- Matters observable by physical inspection which are suggestive of rights of others such as signs, tenancies, prescriptive rights and easements, and adverse possession.[33]

*Alden v. Schuster*[34] illustrates how title companies use the general survey exception to exclude matters that would have been revealed on a survey. In *Alden,* the plaintiffs sought to recover from the insurer under a pre-2006 title policy because improvements from an adjoining property (including a driveway, front walk, wood steps, wooden deck, roof drainage system, and concrete pad) encroached onto their property. Plaintiffs did not obtain a survey when they purchased their property but obtained one upon deciding to build an addition to their house. The title company refused to provide coverage because the policy excluded "any fact which an accurate survey . . . would disclose." In its analysis, the court explained the purpose of such an exception:

> From a search of relevant public records, a title company cannot ascertain the risks that an accurate survey would disclose. It is for this reason that the title company puts that risk on the insured, who can control it either by obtaining a survey or arranging for the elimination of the survey exception. Thus, the very purpose of a survey exception is to exclude

---

survey, with the court considering an accurate survey for purposes of the title insurance in question to mean an "ordinary survey" and citing Friedman on Contracts and Conveyances of Property § 10:6 (7th ed. 2005), at 164, in support of the proposition that an accurate survey does not necessarily mean a full-blown architect's or engineer's survey which would provide additional data pertaining to adjoining properties and thereby identify a possible title defect which would not be shown by a more limited "ordinary survey"), *aff'd,* 874 N.Y.S.2d 857 (4th Dep't 2009).

33. Where a surveyor's legal description references total area of a parcel (i.e., total acreage or square footage), the title policy will normally contain an exception in Schedule B or a note for information that "the exact area or acreage of the premises is not insured." It is the position of the title companies that the calculation of area is exclusively within the abilities and responsibility of the surveyor. Sometimes, title companies will specifically insure acreage for an additional charge where they can rely on sophisticated survey information that sets forth precise acreage data.

34. Alden v. Schuster, No. CV065000475S, 2008 WL 2039075, 45 Conn. L. Rptr. 408 (Conn. Super. Ct. Apr. 22, 2008).

> from coverage errors that would be revealed not by a search of the public records, but by an accurate survey.[35]

The court, therefore, granted summary judgment in favor of the insurer based on the general survey exception in the policy.

Another common issue that arises under the pre-2006 Forms is whether the policy, in the absence of an exception or survey endorsement, extends to encroachments onto adjoining property.[36] Inherent in such a dispute is the issue of whether an encroachment onto adjoining land creates a title defect with respect to the land from which the encroachment emanates. For example, in *First American Title Insurance Co. v. Dahlmann,* the title company argued that the policy did not insure title to any property beyond the bounds described in Schedule A, and, therefore, the title company was not liable for damages resulting from an encroachment of an underground parking garage onto adjoining property.[37] The *Dahlmann* court held that an encroachment onto adjoining property, if substantial, is an encumbrance on title which is covered by the policy, noting that the title company had deleted the standard exceptions for encroachments and survey matters in return for an additional premium to provide "extended coverage" and stating that an argument to the contrary "misconstrues the purpose of title insurance as insuring land rather than insuring title."[38] The case was remanded for a factual determination of whether the encroachment was "substantial."

Determining a policy's coverage with respect to encroachments onto adjoining land is not always as simple as the inclusion or deletion of the survey exception from the policy.[39] For example, in *Ben-Avraham v. Lawyers Title Insurance Co.,* the insured condominium owner looked to the title company to defend an action brought by an adjoining building owner for an encroachment onto that adjoining property's air rights.[40] The title company argued that it was not required to defend the insured because the policy included a standard exception for facts that an accurate survey would reveal.[41] The court noted that "an insurer must establish that the exclusion is stated in clear and unmistakable language, is subject to no other reasonable interpretation, and applies in the particular case."[42] When the policy was issued, the deed recited that the property description was subject to the facts shown on a particular survey, and the title insurance commitment

35. *Id.* at *2 (citing Am. Title Ins. Co. v. Carter, 670 So.2d 1115 (Fla. Dist. Ct. App. 1996) and Walker Rogge, Inc. v. Chelsea Title & Guar. Co., 562 A.2d 208, 212 (N.J. 1989)).

36. Hammann, *supra* note 8.

37. First Am. Title Ins. Co. v. Dahlmann, 715 N.W.2d 609 (Wis. 2006).

38. *Id.* at 176.

39. Ben-Avraham v. Lawyers Title Ins. Corp., 786 N.Y.S.2d 272 (N.Y. Sup. Ct. 2004).

40. *Id.* at 274.

41. *Id.*

42. *Id.* (citing Hotel des Artistes, Inc. v. Gen. Accident Ins. Co. of Am., 775 N.Y.S.2d 262, 268 (1st Dep't 2004) and quoting Continental Casualty Co. v. Rapid-American Corp., 609 N.E.2d 506 (N.Y. 1993)); L. Smirlock Realty Corp. v. Title Guar. Co., 418 N.E.2d 650 (N.Y. 1981), and cases cited therein.

did not include an exception for facts that an additional survey would reveal.[43] In addition, the court noted that "the condominium endorsement expressly insured against 'any present encroachments' without qualification, which is the purpose of title insurance."[44] Here, the court found in favor of the insured, noting that exclusions are to be strictly interpreted.[45] Thus, if title companies do not carefully construct the exceptions to their policies, they may find themselves providing coverage beyond the scope of what they intended to provide.

## *B. 2006 Forms*

As indicated above, the 2006 Forms resolve some of the previous confusion by adding survey coverage as an insured risk in new Covered Risk 2(c), thereby insuring against loss due to "any encroachment, encumbrance, violation, variation, or adverse circumstance affecting the Title that would be disclosed by an accurate and complete land survey of the Land," regardless of whether an accurate and complete survey exists when the policy is issued.[46]

Covered Risk 2(c) of the 2006 Forms further clarifies the extent of coverage by adding a new definition for and insurance against "any encroachment" stating that: "The term 'encroachment' includes encroachments of existing improvements located on the Land onto adjoining land, and encroachments onto the Land of existing improvements located on adjoining land."[47] This definition creates an express recognition that survey coverage extends to losses from improvements that encroach onto adjoining land rather than only those which encroach upon insured land described in Schedule A to the policy.[48]

The 2006 Forms' inclusion of survey coverage language has not, however, totally eliminated the need for surveys to obtain coverage for loss due to matters that would be disclosed by an accurate and complete land survey exception. For example, until November 2008, the New York Endorsement and the Title Insurance Rate Service Association (TIRSA) Rate Manual provided for removal of Covered Risk 2(c) from all New York policies except for loan policies insuring one- to four-family residences.[49] As a result, loan policies for mixed use, commercial, and vacant property, as well as all owners' policies, did not include automatic survey coverage language until the Standard New York Endorsement was revised to eliminate "Item 1" which deleted Covered Risk 2(c).[50]

43. *Ben-Avraham,* 786 N.Y.S.2d at 792.

44. *Id.* at 793–94.

45. *Id.* at 794 (citing Incorporated Village of Cedarhurst v. Hanover Ins. Co., 675 N.E.2d 822 (N.Y. 1996)).

46. Covered Risk 2(c) in 2006 Forms, http://www.alta.org (last visited Nov. 2, 2011).

47. *Id.*

48. Gosdin, *supra* note 3, at 12.

49. LandAmerica Lawyers Title, Underwriting Bulletin No. 13–2008 (2008), http://www.advantagetitle.com/news-and-updates/underwriter-pdf/lalt_13-2008.pdf (last visited Nov. 3, 2011).

50. *Id.*

Although Covered Risk 2(c) in the 2006 Forms provides automatic survey coverage and eliminates the need to find coverage based on the deletion of an exception, this does not prevent title insurers from inserting exceptions for survey matters. The exceptions, generally set forth on Schedule B to the policy, can consist of standard printed exceptions (which differ regionally, as discussed above) and other specific exceptions based on relevant facts. If the insured does not provide an acceptable survey, the title company may include a policy exception for all defects that would have been discovered by a complete and accurate survey.[51] If a survey is provided, the title company may decide not to include a survey exception, or it may elect to include an exception in the policy for any state of facts shown by the referenced survey. If the insured is not able to get this exception removed, it should have the exception revised to "read in" the survey exceptions so that only those specifically described will be excepted from coverage.[52]

Title companies do not warrant the complete accuracy of surveys even when they are "read into" policies, and the title company may be liable to the insured for title defects even if they are not shown by a referenced survey. For example, if a survey fails to show that a building impermissibly encroaches upon adjoining property, the insurer may be liable for damages incurred to remove the encroachment.[53] In such cases, title insurers may seek indemnification from the surveyor for losses arising from the defective survey.[54] Nonetheless, not every inaccuracy on a survey creates an insured title defect. To illustrate, a survey that correctly depicts boundary lines but incorrectly locates the building—thereby leaving the insured to believe that zoning setback lines are not violated—will not present a sustainable claim against the insurer except to the extent that coverage is provided by special affirmative insurance or where there is record notice. This result occurs because, absent special affirmative insurance, losses arising from violations of governmental laws are excluded from coverage unless there is record notice of the violation.[55]

---

51. Gosdin, *supra* note 3, at 135 (The standard survey exception in the owner's policy states: "Rights or claims of parties in possession and easements or claims of easements not shown by the public records, boundary line disputes, overlaps, encroachments, and any matters not of record which would be disclosed by an accurate survey and inspection of the land." The standard survey exception in the loan policy provides: "Any state of facts as would be disclosed by an accurate survey and inspection of the premises.").

52. Marvin N. Bagwell, *New Title Policy Forms Coming to New York,* 21 N.Y. Real Estate L. Rep. 1, 2 (2007).

53. Broadway Realty Co. v. Lawyers Title Ins. & Trust Co., 123 N.E. 754, 757 (N.Y. 1919) ("The exception protects the insurance company from things hidden or done clandestinely, but not from the very acts insured against.").

54. *See, e.g.,* Commonwealth Land Title Ins. Co. v. Close, Jensen & Miller, P.C., No. CV065003046, 2008 WL 4926953 (Conn. Super. Ct. Nov. 5, 2008).

55. Exclusion 1 and Covered Risk 5 in 2006 Forms, http://www.titlelaw-newyork.com/ALTA/2006OwnersPolicy.pdf, http://www.titlelaw-newyork.com/ALTA/2006LoanPolicy.pdf (last visited Nov. 4, 2011).

### C. Endorsements

Survey coverage under Covered Risk 2(c) does not eliminate the need for considering additional coverage available through special affirmative endorsements.[56] Representative affirmative endorsements that may be appropriate include:

- Insurance that the property insured is the same as the property shown on a referenced survey;
- Insurance that multiple parcels are contiguous if so shown by the survey;
- Insurance as to acreage or, if refused by the insurer, a limitation on acreage exceptions with affirmative insurance as to boundary courses and distances;
- Insurance as to specified access (i.e., identified public street or specified easement);
- Insurance as to specific locations of easements and improvements (where blanket utility easements are involved, insurance against loss from, or use of, such easements sometimes may be obtained even though they are not specifically located); and
- Insurance as to no encroachments of improvements by or onto the property, easements, or setback lines (insurance against zoning setback violations may require a zoning opinion).[57]

## IV. Some Final Thoughts

Only time will tell how the detailed listing of specifically covered losses in Covered Risk 2 will ultimately impact title insurance companies and policyholders. The commentators cited in this chapter agree that the listing was made to clarify and expand coverage to the advantage of policyholders and should not be viewed as an exclusive listing of covered losses. But is it necessarily the case that coverage potentially in conflict with specific language will be found to exist for the benefit of the insured?

For example, Covered Risk 2(c) now provides express protection against loss or damage from "[a]ny encroachment, encumbrance, violation, variation, or adverse circumstance affecting the Title that would be disclosed by an accurate and complete land survey of the Land." While an insured can now point to coverage on the basis of a stated provision rather than needing to claim coverage based on the deletion of an exception from coverage for an encroachment that would be disclosed by an accurate and complete survey, is the insured home free for all

56. *See* Gosdin, *supra* note 3, at 13.

57. *See* James L. Lipscomb et al., *Survey Type Coverage Desired by the Institutional Investor, in* Title Insurance 1989: Negotiating Additional Coverages, 331 (PLI Real Est. L. and Practice, Course Handbook Series No. 331, 1989) (discussing survey related endorsements); *see also* Gosdin, *supra* note 3, at 13.

encroachments, etc. absent a survey read into the policy that shows encroachments? Not necessarily.

What if an owner or lender holding a policy that includes Covered Risk 2(c) without exception ends up suffering a loss or damage because of an underground encroachment that is not disclosed by record documents, is not known to the insured, and would not be visible or readily observable by a surveyor who is performing an ALTA/ASCM Land Title Survey in accordance with the Minimum Standards? Will the title company have to pay for such loss or damage? Maybe yes, maybe no—depending on how the court construes the coverage language and whether the court determines that the meaning of "an accurate and complete survey" is to be construed against the insurer so as to cover all encroachments regardless of the standard of performance to be used by the surveyor since the insurer could expressly except coverage for subsurface encroachments. Then again, prior custom and practice has been to obtain affirmative coverage for subsurface encroachments through an affirmative endorsement even if there is a survey that does not disclose subsurface encroachments. It would seem that there is still room for ambiguity on the issue of coverage for encroachments so that insurers and insureds might want to address the issue by clearly written endorsements and/or exceptions, as appropriate.

# 10 The Surveyor as Expert Witness[1]

Robert W. Foster, PE, PLS

When litigation requires the expertise of a professional surveyor, there are things the professional needs to know about the process. For most surveyors, there may be only a few opportunities in a whole career to testify in trial; it is a process some never become fully comfortable with. Not the least of the problems for some surveyors is understanding the role of the expert witness. In order to understand the role, it is necessary to understand the surveyor/attorney relationship since that is usually where the role begins.

## I. The Surveyor/Attorney Relationship

### A. The Fee Arrangement

> There are a number of things that must occur before you ever climb on the witness stand. First of all, and it may be the most important, better have a good fee schedule lined up and know who is going to pay you . . . never under any circumstances have a contingent fee . . . a fee that is based on the outcome of the case. That is a criminal offense . . . an expert witness is supposed to be there to testify to his opinion . . . and to exercise judgment and discretion; and if the outcome of the case is going to affect . . . his remuneration, then he has a situation which

1. An earlier version of this article appeared in *P.O.B.*, March 2000.

> prevents him from being a true expert witness, and his testimony is subject to violent impeachment.[2]

Attorneys understand about fee arrangements and will always want to know what the expert's fee schedule is. In my experience, it is best to be engaged by the attorney, not the attorney's client. The fee payment should come from the attorney, as should an engagement letter that clearly states the intended role of the expert and the agreed-upon fee arrangement. Attorneys are accustomed to making retainer payments to experts, but it is understood that a retainer is not a lump-sum payment for the intended services.

### *B. Case Evaluation*

Before agreeing to testify, the professional must make her own personal evaluation of the merits of the case. The professional witness's first obligation is to advise her client of the merits of the case from the professional's point of view and explain whether she is able to provide supporting testimony in the case. The value of a professional witness depends upon total objectivity. If she is to provide supportive testimony, the witness must first determine that there are merits to be supported.

### *C. Preparation*

The attorney discusses the merits of the case long before the trial date. He wants to know whether the surveyor can testify in an unimpeachable fashion to sustain a position that he is trying to establish, and do so within the framework of the ethics of his profession with honesty, accuracy and knowledge.[3]

There must be a clear understanding by the expert witness and the attorney of their respective roles and the rules under which they must operate. The attorney should brief the witness on the intended procedures to be followed in offering the witness's testimony.

The rules of evidence that may affect the witness's testimony should be explained to the professional witness; the limits imposed on the witness by professional rules and ethics should be explained to the attorney. (It should not be, but often is, necessary to point out to the attorney that the surveyor's first duty is to the court, not to the attorney or the attorney's client.) It is of paramount importance that the attorney know the content of the witness's intended testimony including any weaknesses the witness sees in the case. It is also of paramount importance that the witness have ample time to review the case, read the documents, visit the site, and form a sound opinion.

---

2. Thomas Breckinridge, *The Surveyor in Court as Witness,* 23(4) Surveying & Mapping J. 568 (1963).
3. Donald E. Ward, *The Land Surveyor's Day in Court,* 30(3) Surveying & Mapping J. 377 (1970).

## D. Qualification of the Expert Witness

According to Rule 702 of the Federal Rules of Evidence one qualifies to testify as an expert in the U.S. court system through a demonstration of "knowledge, skill, experience, training or education." A surveyor demonstrates his qualifications through a recitation of these qualifications on the questioning of the attorney. Professional registration or licensure, formal education, continuing education courses, participation in professional associations, articles or books the witness may have written as well as seminars and workshops the witness may have taught, and a description of practical surveying experience are all examples of qualifications a court wants to hear from an expert witness. The witness should be given an opportunity to make a complete review of qualifications. An opposing attorney may interrupt and offer to accept the witness as an expert, not wanting a jury to hear how well the witness is qualified. Another opposing attorney may question the witness on qualifications. In any case, a witness's claim of qualifications should be complete and supportable. Through it all, the attorney for whom the expert is testifying should be fully familiar with all the expert's qualifications and should draw them all out in her questioning of the witness during the qualification process.

## E. The Role of the Expert

According to Rule 702 of the Federal Rules of Evidence, the role of an expert witness is "to assist the trier of fact to understand the evidence, or to determine a fact in issue." The major distinction between an expert witness and a lay witness lies in the ability of the expert to offer an opinion in his testimony. Quoting from Rule 702 of the Federal Rules of Evidence, "a witness qualified as an expert by knowledge, skill, experience, training or education, may testify thereto *in the form of an opinion* or otherwise." (Emphasis added.) It is also well understood that an expert witness is a professional and is to be paid for his or her testimony and preparation, which is another distinction between the expert and lay witness.

## F. Post-Trial Discussions

Attorneys are inclined to walk away from a case after its conclusion with hardly a look backward, but for the expert witness, it is important to review his testimony and its effectiveness—especially if the witness expects to testify in the future. Since a case may continue for days or weeks after an expert has completed testifying, it is often necessary to make an effort to contact the attorney for an evaluation of the witness's testimony and its effect on the outcome of the trial.

The surveyor as expert witness should have no difficulty working with an experienced trial attorney. First comes the engagement letter. Then comes the surveyor's evaluation of the case based on a full review. Finally, having found merit in the attorney's case, the professional surveyor offers testimony and opinion as an expert witness. With the right understanding between them, and with thorough preparation by the witness, the surveyor/lawyer relationship can be a smooth and satisfying one—professionally speaking.

## II. Intra-Professional Testimony

The role of the surveyor as expert, when the subject is a negligence suit against another professional surveyor, is one that many surveyors will avoid. We know that the surveyor is often called upon to testify in a land dispute, and the surveyor may in that case testify in support of her own survey or in response to a survey by someone else. The more problematic situation comes when the surveyor is asked to testify on behalf of the plaintiff in a tort case in which another professional surveyor has been charged with negligence or breach of duty.

Some in the profession feel that testimony for the plaintiff in an action against another surveyor would be an act of treachery against a fellow professional or an attack on the integrity of the profession. It is also true that such testimony may create enmity between professionals and is not likely to enhance friendships.

The answer to all these objections is another question: if a surveyor is unwilling to appear in court to explain the duty of care of the professional surveyor and to testify as to the failure of a defendant to meet that duty, who will define these things for the court? Lawyers and judges do not ordinarily have a clear understanding of the standards of care expected in the field of surveying. People do make mistakes; people are injured; and lawyers do bring charges. Someone must be willing to take a position in order to clarify the issues for the court. Some will recommend leaving it to the boards of licensure to punish the malefactors in the profession. But a property owner who has suffered loss because of a poorly executed survey will be looking for damages through tort litigation and will be less interested in bringing charges before the licensing board which provides no recompense beyond willful satisfaction. Finally, we should remember that licensure is a privilege granted by the state on the presumption that one's first duty is to the public. Professionals have a responsibility to police their own profession. The profession is quick to congratulate noteworthy members with awards; it must be willing to support charges against the blameworthy as well.

Needless to say, it is a much more comfortable position to provide expert testimony in *defense* of another professional charged with negligence or breach of duty. The same rule applies: the court requires definition of the standards of care in surveying and needs someone to give expert testimony to show that the defendant surveyor met his or her duty. (The professional charged with negligence cannot ordinarily testify as an expert in his or her own defense because of a lack of objectivity.)

Whether testifying for the plaintiff or the defendant, the same process is required of the professional witness. First, she must enter into an agreement of services with the attorney who hires her to testify. (An expert should be employed by the attorney, not the attorney's client.) They will agree to terms and execute a letter of agreement. Next, the expert will make a case evaluation, studying the charges, the defendant's answers, and all the documents relating to the case—before agreeing to testify. Satisfied that she can give unbiased and knowledgeable testimony, she notifies the attorney of her willingness to proceed.

This application of forensics in surveying is a necessary and honorable task, whether testifying on behalf of a plaintiff or a defendant in a negligence case. In fact, one's credibility in the courtroom may be enhanced for the witness with a record of testifying for plaintiffs *and* defendants. It is not a good idea to become known exclusively as either a plaintiff's witness or a defendant's witness; juries are inclined to hold in higher respect the unbiased professional who is willing to stand on either side of the charges and let the chips fall where they may.

As for the "hired gun" accusation, it is an overrated insult. The courts are well aware of the role of the expert witness and are quite willing to explain their value to a jury. A cross-examining attorney who had reviewed my history of professional testimony once directed the epithet at me. When he suggested I was a hired gun, the attorneys for whom I was testifying exploded, the judge came out of his chair, and several jurors showed expressions of disgust. In post-trial conferences, we learned from one juror that the whole panel was disgusted at that attorney's attempt to degrade my testimony; perhaps they saw it as an insult to their own intelligence. In any case, I have learned not to worry about the tactics and antics of cross-examining attorneys. They are doing their jobs as advocates; I am on solid ground when I am totally objective and well prepared to do my job as a professional witness—but I have to be willing to serve in that position in the first place. I recommend the practice to all qualified professionals with a concern for their profession—and with a reasonably thick skin!

## III. Expert Witness Immunity: Is There Any?

Professional surveyors are often called to testify as expert witnesses. They are (or should be) well compensated for their services, which include not only time in court, but also time in preparation to deliver testimony. Preparation time is often far in excess of actual time spent on the stand and may include strategizing a case with an attorney as well as the professional's preparing himself for testimony. Indeed, preparation may include advice given by the expert to the lawyer. The lawyer may later employ the expert's advice in settling a case, in which instance the witness may never take the stand. In short, an expert's participation may go well beyond actual court testimony given during trial or testimony given in deposition. The foregoing are all well-recognized professional services. The question is to what extent may an expert be held liable when charged with a failure to meet her duty of care?

Guidelines for Improving Practice[4] carried an article titled *Expanding Risk for Expert Witnesses,* in which the witness immunity doctrine was discussed: "Expert witnesses are considered to be *amici,* or friends of the court—professionals who willingly take time to serve the cause of justice . . . Because of their special role in the legal system, expert witnesses have heretofore been entitled to absolute immunity from civil liability for anything they said on the witness stand." The article goes on to point out that witnesses do more than testify. They investigate failure, do research and investigation, provide advice, provide most services outside the courtroom, and are seen increasingly as "hired guns rather than as *amici.*"

Then comes the key statement in the Guidelines article: "Expert witnesses owe their clients certain duties by virtue of their specialized knowledge, skills or training which, when breached, can be grounds for a cause of action by a client who has been damaged." Schinnerer, of course, is in the professional liability insurance business, has decades of experience working with engineers, architects, and surveyors, and is understandably concerned about any expansion of liability affecting its policyholders. In the article, Schinnerer reports the differing actions of courts in several states in dealing with expert witness immunity. The following hypothetical cases are an attempt to translate into the surveyor's purview the significance of these decisions.

Let's say you are appointed by the court to consider a land boundary dispute between two property owners. Both parties in the case agree that your opinion will be binding. You investigate the case and conclude that the surveyor for Party A has made a correct determination of the property line and that the line set by the surveyor for Party B should be disregarded. The court rules, holding to your conclusion. In the aftermath of the case, the surveyor for Party B convinces his client that you made an error in your determination in that you favored a record description over evidence found. Party B sues you. Your defense is that you should receive judicial immunity since you acted as a court-appointed arbitrator in the case. Schinnerer reports that New Jersey appears to be the only state in which a court has held that not even a court-appointed expert is immune from liability for deviating from the applicable accepted professional standards, and makes reference to *Levine v. Wiss & Co.,* in which the court stated that the standard of reasonable care used for most professionals was applicable.[5]

In another scenario, imagine a dispute between a contractor and a public agency over the amount of fill required to complete a project. You were hired by the contractor to determine the amount of fill. You make a topographic survey, make calculations, and testify accordingly. It is determined later that either your survey was inaccurate or your calculations were incorrect. You are sued for negligence in not meeting your duty of care. You plead immunity as an expert. A 1989 case in the state of Washington[6] had a different outcome than the one in New

4. Victor O. Schinnerer & Co., 31(1) Guidelines for Improving Practice.

5. Levine v. Wiss & Co., 478 A.2d 397 (N. J. 1984).

6. Bruce v. Bryne-Stevens & Assocs. Eng'rs, Inc., 776 P.2d 666 (Wash. 1989).

Jersey. Schinnerer reports that the Washington state supreme court determined that the fact that the engineer had been retained and compensated by the plaintiff, rather than appointed by the court, did not deprive him of expert witness immunity. The court also held that witness immunity applied not only to the engineer's testimony, but also to any action the engineer took before trial that helped form the basis of his testimony.

The outcome might have been different again in California where, according to Schinnerer, in *Mattco Forge, Inc. v. Arthur Young & Co.*,[7] the court said that the witness immunity doctrine was not established to protect a litigant's own expert, but to protect experts from harassment claims by adverse litigants. In words no witness wants to hear, the court said that applying the privilege "does not encourage witnesses to testify truthfully; indeed, by shielding a negligent expert witness from liability, it has the opposite effect."[8]

The Missouri Supreme Court seems to agree with the California appeals court. In *Murphy v. A.A. Mathews*,[9] Schinnerer explains that the court held that privately retained professionals who negligently provide litigation-related services should not be covered by witness immunity. That protection, the *Murphy* court said, should only cover defamation suits and retaliatory actions against adverse witnesses.

There are several categories in which an expert might wonder about his exposure to liability claims: (1) when he acts as a "friendly witness," appointed by the court, (2) when negligence is claimed by his client for whom he has testified, (3) when he is sued by the opposing party in a case in which he testified for a litigant, and (4) when he is sued for negligence in preparation of documentation used in litigation. The outcome might depend upon which category applies and upon the state in which the expert provided the service. But don't be too sure of anything. The Schinnerer article concludes: "Firms that provide these services should know that while the essential status of the expert witness remains the same, the times—and the risk—may be changing indeed."

7. 6 Cal. Rptr. 2d 781 (Cal Ct. App. 1992).
8. *Id.* at 782.
9. 841 S.W.2d 671 (Mo. 1992).

# The Surveyor's Role in Boundary Disputes[1]

11

Robert W. Foster, PE, PLS

## I. The Surveyor's Role

The role of the surveyor was defined famously by Thomas M. Cooley, Chief Justice of the Supreme Court of the State of Michigan in an essay first published in the Michigan Engineering Society Journal late in the nineteenth century.[2] In *The Judicial Function of Surveyors,* Judge Cooley offered two rules for the surveyor when property boundaries are in dispute. The first rule is "to search for original monuments or the place where they were originally located. . . ." The second rule is, when original monuments are no longer discoverable, "the question of location becomes one of evidence merely." State statutes should not direct the surveyor "to locate or establish a corner, as the place of the original monument, according to some inflexible rule." Instead the surveyor "must inquire into all the facts . . . always keeping in mind, first, *that neither his opinion nor his survey can be conclusive upon parties concerned,* and, second, that courts and juries may be required to follow after the surveyor over the same ground, and that it is exceedingly desirable that he govern his action by the same lights and the same rules that will govern theirs." (emphasis added). Judge Cooley summarizes by observing that "(s)urveyors are not and

1. An earlier version of this article appeared in P.O.B., July 2005.

2. This essay was reprinted in the July–Sept. 1953 issue of SURVEYING & MAPPING, a journal of the American Congress on Surveying and Mapping, and the *The Surveyor in Court,* an ACSM-published collection of articles first appearing in SURVEYING & MAPPING.

cannot be judicial officers, but in a great many cases they act in a quasi-judicial capacity with the acquiescence of the parties concerned. . . ." To put it another way, the culmination of the surveyor's work is an opinion expressed by the plan he produces and by the marks or monuments he sets on the ground. He has no official judicial function or authority, but his opinion will be accepted most of the time by the property owners affected until or unless it is overturned by a court.

## II. Surveyor Judgment

It is clear that there are several principles for the surveyor to consider before arriving at an opinion as to the location of a property boundary. In construing a deed, she must consider the superiority of calls for monuments over distance and direction and direction over distance except where distance would be more reliable—but she must first be satisfied that the monuments she has recovered in her field survey are the original called-for monuments. She must attempt to find original intent of the grantor and grantee when the description is ambiguous or inadequate. She must also consider both the general and the particular descriptions of a deed and make a judgment when there is a contradiction between the two. Another principle is that any ambiguity in the description should be resolved in favor of the grantee. All these judgments require careful study and thoughtful consideration, as they may well be contradicted by another surveyor and may be overturned as a result of litigation. Williams and Onsrud explain it this way: "If a court upheld the surveyor's evaluation of the evidence in the example, it is because the surveyor arrived at a comprehensive and well-reasoned answer rather than because he arrived at the theoretically correct answer. . . . [T]here are no 'true' answers waiting to be discovered, only well-reasoned answers."[3]

## III. Professional Cooperation

Boundary disputes occur when owners of adjacent land disagree on the location of the line between their properties. This may be because of their own impressions of where the line should be, but more often it is because a surveyor for one of the owners has defined the line contrary to the understanding or belief of the other owner. If the second owner cannot be convinced by the first neighbor's surveyor, he will hire his own surveyor, hoping to support his own understanding or belief. Logic suggests that the second surveyor will come to the same conclusion as the first surveyor as to the location of the line. But history and experience show otherwise. Each surveyor will come to her own conclusions based on a study of the deeds involved and the

3. Mitchell G. Williams & Harlan J. Onsrud, *What Every Lawyer Should Know About Title Surveys, in* Land Surveys: A Guide for Lawyers and Other Professionals (2d ed. 2003).

conditions on the ground. When there are ambiguities, contradictions, or errors in the record or as evidenced in conditions on the ground, surveyors must weigh the evidence and arrive at their own conclusions. Williams and Onsrud explain that "[s]urveyors occasionally disagree on the proper location of a boundary line not necessarily because one surveyor measures better than the other, but more commonly because each surveyor has weighed the evidence differently and has formed different opinions."[4]

Two surveyors coming to differing "well-reasoned answers" have a regulatory obligation in some jurisdictions to seek a common understanding. For instance: "In the event of substantial disagreement with the work of another surveyor, contact the other surveyor and investigate the disagreement."[5] Besides a regulatory obligation, surveyors have an ethical obligation to reconcile their differences in order to meet their social responsibilities. "In the event a property surveyor . . . disagrees with the work of another property surveyor, it is the duty of that surveyor to inform the other surveyor of such fact."[6] By following a regulatory directive and by recognizing their professional duty, two surveyors with differing conclusions in their neighboring surveys are in a position to avoid or settle a dispute between abutting property owners—a dispute that could lead to expensive and protracted litigation. Unfortunately, a misplaced concern for their clients' interests, ego, and a reluctance to yield to another's opinion or a concern for supposed professional liability exposure often mitigate against cooperation between surveyors with contradictory opinions. All too often the result is litigation wherein a court will arrive at its own conclusions regarding the disputed boundary.

Each professional surveyor has a duty to his client and an equal duty to his client's neighbor. Both surveyors ought to recognize they are in a position to keep peace in the neighborhood—if only they can reach a collective "well-reasoned answer."

## IV. Summary

In summation I offer the following six rules for the surveyor in performance of boundary surveying:

Rule 1: The quasi-judicial capacity of surveyors. Thomas M. Cooley, Chief Justice of the Supreme Court of Michigan in 1953, summarized the role of the surveyor: "Surveyors are not and cannot be judicial officers, but in a great many cases they act in a quasi-judicial capacity with the acquiescence

4. *Id.*

5. 250 Mass. Code Regs. 6.01(1)(d)2, "Procedural and Technical Standards for the Practice of Land Surveying."

6. Curtis M. Brown, Walter G. Robillard, & Donald A. Wilson, Evidence and Procedures for Boundary Location p. 520 (6th ed. 2011).

of the parties concerned . . ." In other words, when the decision of the professional surveyor is accepted by all interested parties, it is as though a judicial decision had been rendered.

Rule 2: The surveyor's responsibility to the public is equal to his responsibility to his client. Whether in performing a retracement survey or in attempting to reconcile a boundary dispute, the surveyor has an obligation to the public (i.e., his client's neighbor) equal to his responsibility to his client's interests. It is a situation endemic to the professional status assigned to the surveyor through public licensure.

Rule 3: Surveyor objectivity. The surveyor and the attorney have distinctly different roles when litigating a boundary dispute. The attorney is an advocate for her client, while the surveyor is a witness offering expert testimony in an unbiased, objective manner.

Rule 4: No "true" answers are waiting to be discovered, only well-reasoned answers. "If a court upheld the surveyor's evaluation of the evidence in the example, it is because the surveyor arrived at a comprehensive and well-reasoned answer rather than because he arrived at the theoretically correct answer. . . . [T]here are no "true" answers waiting to be discovered, only well-reasoned answers."[7]

Rule 5: Professional cooperation between surveyors. Two surveyors coming to differing "well-reasoned answers" have a regulatory obligation in some jurisdictions to seek a common understanding and, in any case, an ethical obligation to reconcile their differences in order to meet their social responsibilities and to keep peace in the neighborhood.

Rule 6: Keep peace in the neighborhood. The physician's oath to do no harm may be reflected in the surveyor's ethical obligation not to disrupt peace in the neighborhood. The source of boundary disputes between neighbors is often found within the realm of the surveyor's expertise, and it is often within the surveyor's ability to reconcile differences. It is also true that many boundary disputes have their origin in the retracement work of a surveyor who, in his zeal to serve his client's interests, has failed to recognize a neighbor's just claim.

7. Williams & Onsrud, *supra* note 3.

# 12 Surveying's Legal Body of Knowledge

Wendy Lathrop, PLS, CFM

It is an all too familiar scenario: a surveyor serving as an expert witness cites a legal resource to support his or her opinion, and the opposing attorney objects, accusing the surveyor of practicing law. However, the knowledge of a body of law and its application to the location of the boundary lines of a parcel of land or an easement, or interpreting a deed description is not the practice of law, but the practice of land surveying. Surveyors, like accountants, need to have an intimate knowledge of the law applicable to their respective professions. Most state licensing boards require applicants for licensure as surveyors to demonstrate that they have that knowledge through education and/or by examination.

The practice of every design professional—including surveyors, geometricians, or whatever else we choose to call ourselves—is intimately tied to the law. To begin with, the very scope of legal and proper practice is defined by statute in each of the individual states, commonwealths, and territories that make up our nation (as is the case in much of the rest of the global community). There is no better way to be sure one is doing all that is permitted and required by law and avoiding all that would constitute malpractice or negligence than to find and read the laws defining practice and misconduct.

Surveyors may be called to court as defendants, professional witnesses, or expert witnesses. Attorneys are educated in generalities and therefore often must rely upon specialists to assist them in specific areas. Surveyors may then be called upon as "expert witnesses," to comment on the facts, but not on the issues of laws. That part is reserved for the judge and jury to determine. It is the surveyor's responsibility as an expert to guide attorneys to the most useful laws and cases, and to suggest lines of questioning to them that may assist in drawing out the facts of a case. A thorough command of the law and associated rulings positions the surveying professional to fulfill this courtroom role.

The body of knowledge for surveyors therefore encompasses a range of legal principles and laws that guide professional, business, and technical practices. A wide variety of legal materials provides this necessary information, and expert surveyors access them both through the Internet and in hard copy at law libraries.

Beginning with the most basic of required legal knowledge, our first example of the importance of finding and reading the law answers the question of what it means to be a surveyor in each of the jurisdictions that regulate the profession. While significantly important to those licensed in a single state, commonwealth, or territory, those licensed in multiple jurisdictions must be careful to stay within the defined limits of practice and be proficient to the fullest extent of those limits.

What may be permitted under one surveying license may not be permitted under another. Maryland, for example, specifically allows (and expects) licensed professional land surveyors to prepare and design plans that include "road and street grades; sediment and erosion control measures; nonpressurized closed storm drainage and stormwater management systems; and open conduit storm drainage and stormwater management systems" when such work is "undertaken in conjunction with site development or subdivision."[1] In nearby New Jersey, however, such design work is beyond what a professional land surveyor may undertake[2] and is encompassed within licensed engineering. Performance of such tasks without an engineering license is termed "unlawful"[3] and is subject to penalties and civil actions by the State Board of Professional Engineers and Land Surveyors.[4]

Included in the defined scope of services that surveyors may perform within the states, commonwealths, and territories is an expectation that the licensed professional be familiar with laws relevant to practice in that locale. Even with dramatically different jurisdictional scopes of practice, as with the two states mentioned above, safeguarding the life, health, property, and well-being of the public is the universal purpose of licensure, and such promotion of public welfare is included in the knowledge that the surveyor must have.

In New Jersey, the term "land surveyor" means a "person who is a professional specialist in the technique of measuring land, educated in the principles of mathematics, the related physical and applied sciences, and *the relevant requirements*

1. Md. Code Ann., Bus Occ. & Prof. § 15-101(j)(1)(vi).
2. N.J. Stat. Ann. § 45:8-28(e).
3. N.J. Stat. Ann. § 45:8-27.
4. N.J. Stat. Ann. § 45:8-39.

*of law,* all requisite to the practice of land surveying as attested by his license as a land surveyor."[5]

In Maryland, "'Practice land surveying' means any service, work, documentation, or practice, the performance or preparation of which requires the application of special knowledge of the principles of mathematics, the related physical and applied sciences, and *the requirements of the relevant law,* as applied to: [list of defined activities follows]."[6]

A familiarity with and understanding of "relevant law" is not specific to these two eastern states that happen to lie in close proximity to one another. Texas, originally under Mexican and Spanish rule before gaining statehood and incorporating traces of those earlier legal systems into its own, includes knowledge of relevant laws in the defined scope of professional surveying.[7] Even farther west, Alaska defines the practice of surveying as incorporating "any service or work the adequate performance of which involves the application of special knowledge of the principles of mathematics, the related physical and applied sciences, and the relevant requirements of law for adequate evidence of the act of measuring and locating land. . . ."[8]

Even in states not specifically including similar language as to the body of knowledge necessary for professional surveying practitioners, the mandate to know the law is reflected in the surveyor's duty as a fact finder of where boundaries are located on the ground. Unbiased, objective weighing of facts is central to the professional's responsibility to the public that relies on the license holder's expertise. The distinction between what a boundary is and where it is located has been reiterated in numerous judicial decisions, often phrased similarly to this statement by the 1960 Pennsylvania Superior Court: "The question of what is a boundary line is a matter of law, but the question of where a boundary line, or a corner, is actually located is a question of fact."[9]

The Supreme Court of the United States weighed in on the distinction between determinations of law and of fact as early as 1799: "It is agreed, that if a Jury collect the contents of a deed, and find them, and then find the deed, in haec verba, the Court must regard the deed itself, and not the construction; because, the Jury are not to judge of the law; and the very circumstance of their finding the verdict specially, shows that they disclaim judging of the law, and submit it to the Court."[10]

5. N.J. Stat. Ann. § 45:8–28 (d) (emphasis added).
6. Md. Code Ann., Bus Occ. & Prof. § 15–101(j)(1) (emphasis added).
7. Tex. Occ. Code Ann. § 1071.002(6)(A).
8. Alaska Stat. § 08.48.341(13).
9. Walleigh v. Emery, 163 A.2d 665, 668 (Pa. Super. Ct. 1960). See also *Miles Land Co. v. Hudson Coal Co.*, 91 A. 1061, 1063 (Pa. 1914) for an expansion on this topic.
10. Sims Lessee v. Irvine, 3 U.S. 425 (1799).

Clearly, while a surveyor is not to make legal determinations (unless he or she is also admitted to the bar), the professional must be familiar with how the law treats determinations of fact, which both the surveyor and the jury must weigh. The legal body of knowledge provides a firm basis for the surveyor's career-long pursuit of the broadest possible array of legal information to be applied in the determination of that fact.

What we do (and do not do) as licensed professionals and how we do it is bound up in a variety of legal materials. Let us examine just one application of surveying: land development. Before taking on work in this arena, legal material surveyors must know and understand how to address the following aspects of a project:

- The limit of professional services that a surveyor can provide in the state where the development is to be accomplished.
- Appropriate contractual language to define the surveyor's responsibilities and to limit liabilities.
- Standards of accuracy necessary in measurements for the project.
- The basis of orientation (state plane coordinates required?).
- Significance of deeded land use restrictions.
- Environmental land use restrictions: floodplains, limestone, tidelands, steep slopes, wetlands, other protected areas.
- Acceptable zoned land use.
- Minimum lot size, setbacks, building size and height, and allowable impervious cover.
- The local process and requirements applicable for special permits, variances, etc.
- Water district restrictions and Coastal Management Areas.
- Road access—existing and reserved rights.
- Setbacks to protect solar easement rights of adjoiners.
- The format of final deliverables to the approving authority (paper, digital).
- The balance of all the various regulatory issues, with all that the professional is required to do both legally and ethically.

This is by no means an exhaustive list. But it illustrates the range of professional, technical, and business questions that only legal resources can answer. The law affects how surveyors interact with clients, other surveyors, and the various agencies with which they work. It establishes standards, both technical and professional, to which they must practice. The law more broadly establishes standards for business practice, not specific to surveying. Table 1 illustrates a small sampling of areas in which the law affects surveying.

**Table 1. Legal guidance within three main areas of surveying practice**

| Professionalism | Business practice | Technical practice |
|---|---|---|
| • Defined scope of professional surveying practice<br>• Standard of care<br>• Ethical standards<br>• Definitions of malpractice and misconduct<br>• Rules of evidence regarding expert testimony | • Business formation<br>• Tax schedules<br>• Employee relations (including EEO and ADA requirements)<br>• OSHA and various other health and safety regulations<br>• Contracts and agreements<br>• Torts<br>• Rules of evidence relating to record keeping<br>• Mechanic's liens<br>• Small claims court actions | • State plane coordinate system<br>• Calibration of instruments<br>• Contents of maps and plats<br>• Monuments—acceptable materials and dimensions, where to set them, recording their location<br>• Filing maps, deeds, and other records<br>• Rules of evidence regarding boundaries and monuments<br>• Compiling and applying the various elements of conveyancing |

The guidance in Table 1 is not found in a single location. Instead, it is scattered throughout a variety of legislative, administrative, judicial, and executive documents. These are further divided along jurisdictional lines, from municipal, to county, to state, to federal. To find the materials necessary even for the simplest of projects, the surveyor must know first that such legal guidance exists and must be adhered to, and next where that guidance can be found. A brief summary of materials surveyors should be familiar with is presented in Table 2. The divisions between the four branches of government are another bit of legal knowledge that surveyors must be familiar with in order to find the desired material.

**Table 2. Types of legal materials surveyors should be familiar with**

| Issuing branch of government | Materials included | Areas of surveying affected |
|---|---|---|
| Legislative | Local ordinances<br>State statutes<br>Federal statutes | Professionalism<br>Business practice<br>Technical practice |
| Administrative | State regulations<br>Federal regulations | Professionalism<br>Business practice<br>Technical practice |

**Table 2. Types of legal materials surveyors should be familiar with (*continued*)**

| | | |
|---|---|---|
| Judicial | Rules of evidence<br>County court actions<br>Probate court actions<br>State court actions<br>Federal court actions | Professionalism<br>Business practice |
| Executive | Executive orders (issued by President and governors) | Business practice<br>Technical practice |

Executive orders are not often recognized as affecting surveying. However, such orders can establish procedures, purpose, focus of an agency's agenda, and more. One particular executive order set policy regarding protection of floodplains that has guided federal agency decisions since its proclamation in 1977. President Carter's Executive Order 11988[11] tied together the National Environmental Policy Act of 1969 (as amended), the National Flood Insurance Act of 1968 (as amended), and the Flood Disaster Protection Act of 1973 to "avoid to the extent possible the long and short term adverse impacts associated with the occupancy and modification of floodplains and to avoid direct or indirect support of floodplain development wherever there is a practicable alternative."

This order affects every federal agency's practice and procedures, which in turn colors the rules and regulations issued and administered by those agencies, with results affecting every individual interacting with any of those agencies. The mission statement embodied in this executive order affects floodplain protection and management throughout the nation. Understanding that mission can be key to understanding the "why" of agency actions, with the result of easier interaction and compliance with regulatory requirements.

Ordinances, statutes, and regulations are the most readily identified legal materials affecting surveyors and surveying practice. Without this basis of knowledge, surveyors cannot uphold the duty to protect the health, safety, and welfare of the public as their licenses require. But the other resources listed in Table 2 significantly affect surveying as well. Table 3 presents a small and nonexhaustive sample of the guidance offered through these resources.

**Table 3. Contents of various legal resources**

| Legislative material (statutes and ordinances) | Administrative rules and regulations | Judicial sources |
|---|---|---|
| Definitions of terms<br>Definition of legal practice<br>Standards of practice<br>Elements of contracts<br>Legislative intent<br>Business standards | Standards of practice<br>Business standards<br>Environmental practice<br>Land use | Rules of evidence<br>Explanation of terms<br>Explanations of what laws mean (interpretation by application to certain fact scenarios) |

11. Exec. Order No. 11988, 42 Fed. Reg. 26951 (May 24, 1977).

The judicial sources are the least used by surveyors except when they find themselves in a courtroom, whether as a defendant or witness. At that point, the most usual response is to run to the court reports to find a decision that may have relevance to the current situation, not often easy due to variations in the factual scenarios. Under less stressful conditions, judicial rulings may also provide interpretation of the meaning and appropriate application of statutes.

But even in daily business practice, the court Rules of Evidence provide the basis for developing sound record keeping that can stand up against a dissatisfied client's suit. Federal Rule of Evidence 403 (adopted in similar form by every court in the nation) establishes that documented habit and routine practice can stand as evidence acceptable to the court that conduct of individuals and the business entity was in conformance with that habit and routine. There is much everyday protection for practitioners available throughout the Rules of Evidence, a resource with which every surveyor should be familiar beyond the rules that outline experts' qualifications and limits of their testimony.

Both the range and depth of surveying are rooted in legal material. A curriculum preparing future surveyors must include familiarity with the legal system, what materials are available, and where to find them. Legal knowledge is not as ephemeral as the technology surveyors employ; the principles and very heart of surveying is legal in nature—"Where is the boundary?" "What activities may I legally undertake?" "What standards apply to this project?"—with mathematical and scientific components supporting and supplementing the mental and logical process of applying the law.

## I. Concluding Remarks

Licensed surveyors are expected to know and understand the legal underpinnings of their profession, and licensing examinations in some states focus entirely on the legal aspects of surveying, ethics, and business practice. Without legal knowledge, one cannot become a licensed surveying professional. Without that knowledge and the ability to apply it practically, one cannot fulfill the legally imposed responsibility to protect the public's interests.

# Surveying Services

# 13

Knud E. Hermansen[1]
PhD, PLS, PE, Esq.

## I. Introduction

In an ideal setting, prospective buyers or landowners seeking to improve their property will retain the services of a competent surveyor who will prepare a detailed boundary retracement survey plan. The plan will be duly forwarded by the prospective buyer or landowner to a competent title attorney who will examine both the plan and the title records. A title opinion will be prepared by the attorney, and the buyer or landowner will be advised on the marketability of the title with recommendations for problems that are discovered.

In practice, most landowners seek neither the advice of an attorney or surveyor before making improvements. Most conveyances are handled by an attorney who does not have a current survey plan. Consequently, the attorney relies almost entirely on a limited research of recorded deeds within the client's chain of title to form an opinion on the title. Potential title problems caused by vague boundaries, encroachments, overlaps, and so forth are seldom discovered from examining the deeds alone. Furthermore, the attorney often lacks the skills, knowledge, training, or experience that will allow him or her to identify potential survey problems from a limited examination of the

1. Knud Hermansen has a PhD in civil engineering from Pennsylvania State University and a JD from West Virginia University. He is a professor in civil engineering technology and surveying engineering at the University of Maine and operates a consulting firm offering surveying, engineering, and legal services.

descriptions found in the title abstract. This article is intended to aid the attorney in evaluating surveys and survey information.

## II. Postulates

To understand and evaluate surveys and survey information, certain basic postulates regarding surveys and survey information must be introduced and explained.

### *A. Descriptions and Title*

A good description is not a guarantee of a correct survey or good title. The best description for locating property boundaries is a metes-and-bounds description. However, as the following text and figure show, a seemingly adequate metes-and-bounds description may cause or hide title problems.[2]

Example:

Left Parcel: " . . . Beginning at a 2" X 2" concrete monument set; thence on a true bearing, North 89° 58' 30" East a distance of 280.00 feet to a #4 rebar set; thence South 00° 01' 30" East 230.00 feet to a #4 rebar set; thence South 89° 58' 30" West 280.00 feet to a 2" X 2" concrete monument set; thence North 00° 01' 30" West 230.00 feet to the point of beginning. Containing 1.48 acres.

Right Parcel: " . . . Beginning at a 2" X 2" concrete monument set; thence on a true bearing, South 89° 58' 30" West a distance of 280.00 feet to a #4 rebar set; thence South 00° 01' 30" East 230.00 feet to a #4 rebar set; thence North 89° 58' 30" East 280.00 feet to a 2" X 2" concrete monument set; thence North 00° 01' 30" West 230.00 feet to the point of beginning. Containing 1.48 acres."

2. The problem illustrated by the descriptions and figure could have been avoided if a careful survey had been obtained or if the description had been originally written to convey the East 1/2 and West 1/2 with the common boundary described with certainty.

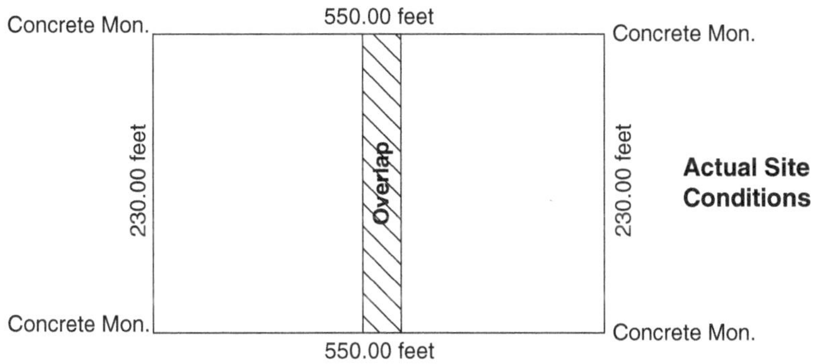


**Figure 13.1** The descriptions describe two parcels, side-by-side, each 280 feet wide. However, the parent tract from which both parcels derive their title is only 550 feet wide. As a result there is 10 feet that is included in both deeds.

Likewise, a bad description is not the same as bad title (i.e., an incomplete or problematic description may convey good title). A bad description may and often does convey good and marketable title, although not adequate to locate the property boundaries without relying on extrinsic evidence.

> Example: "Bounded on the north by the Penobscot River, on the east by Route 2, on the south by James Finnigan's land, and on the west by Jerimiah Ladder's land."

There are many times that a bad description is often a symptom or indication of problematic title. Consequently, the question arises: how to determine if a bad description will cause problems or whether a seemingly adequate description may be hiding problems?

## *B. Adequate in the Past, Inadequate for the Present*

The majority of descriptions currently used were originally written from a survey performed when the property was only worth a fraction of the parcel's present value. In the past, the cost of the survey may have exceeded the value of the property (hence, errors discovered in the survey were seldom corrected). Furthermore, the property was often surveyed and the description written by a landowner, attorney, or surveyor lacking adequate training and using only rudimentary equipment and procedures in comparison to today's standards.

> In closing this report, it may not be improper to call attention to the fact that the various litigations and disputes about boundaries, which our courts of justice are constantly called upon to decide, are most of them either directly or indirectly the result of the present loose and imperfect method of conducting land surveys. This evil is not, however, it must be acknowledged, confined exclusively to the surveyors. Many of our lawyers, who are entrusted with the drafting of instruments of conveyance, are often deficient

> in the knowledge requisite to render their descriptions of land correct and to place them beyond the possibility of a misconstruction.[3]

In cases where the descriptions are adequate, the disappearance of the monuments and marks often leads to the same problems that poor descriptions do—uncertainty, encroachments, and conflict. In some situations, the original corner monuments, if there were any, no longer exist. This situation is often the result of inexpensive previous surveys, generally of poor quality. In other cases where quality surveys were performed, lack of care and attention has resulted in the decay, destruction, and disappearance of the monuments and original marks.

> The deeds to these respective properties in the chain of title are as we so often see . . . inadequate, inartfully drawn and in essence, ineffective in order to attempt to locate and define a boundary line under today's standards. It's a known fact that in days gone by, that landowners were not very particular with their descriptions with their properties. Most went upon the fact that they knew where their property line was, their neighbor knew where it was, and when it came to making deeds, it didn't make much difference what they put in the deed or how they described it. The buyer under those deeds was usually a person local in the area and he knew where it was anyway. And, for that reason . . . have caused all of this litigation that the Court has seen, trying to straighten out boundary lines, and the locations of various properties. It certainly goes without saying that in days gone by, the methods and instruments to be used for making surveys, and defining the boundaries of land, had not been refined to what they are in this day in time, and the results are simply that the old deeds are totally and wholly inadequate to base reasonable boundary line determinations on without resorting to the best abilities that a particular individual that is called upon and trained in that field, and to go upon that property and do the best he can."[4]
>
> The inaccuracy of early surveys, the fine disregard which our pioneers showed for the accurate measurement of their holdings, and the difficulties inherent in the subject all result in boundary line disputes, which become more frequent as the subject matter becomes more valuable.[5]

The question, of course, is how long ago is too long and how does an attorney determine the age of the previous survey without doing all the research and fieldwork the attorney is attempting to avoid, if unwarranted?

---

3. Variations of the Magnetic Needle: Report of the Commissioner on the Variations of the Magnetic Needle, State of Maine 74 (1866).

4. Dickinson v. Bain, No. 03A01–9105CH00163, 1991 WL 169184 (Tenn. Ct. App. Sept. 5, 1991); *see also* Young v. Blakeman, 95 P. 888 (Cal. 1908).

5. Joseph P. Loeb, *The Establishment of Boundary Lines by Practical Location*, 4 Cal. L. Rev. 179, (1916); *see also The Conveyance*, 187 Law Times 258 (1939); Samuel C. Wiel, *Built-Up Boundaries Outweigh Paper Boundaries*, 4Cal. L. Rev. 293 (1916); *Old Law Suits and Old Monuments, Stones, and Lore*, 105 Law Times 427 (1898).

### C. A Surveyor Provides an Opinion, Not a Guarantee

The surveyor is a licensed professional who is sought to provide an opinion and not a product.[6] Surveyors are no different from attorneys, doctors, ministers, and others who offer services to the public. An attorney cannot guarantee a person's innocence or guilt before or after a trial; a doctor cannot guarantee a patient's perfect health, cure for a malady, or demise after a thorough physical exam or medical attention; and a minister cannot guarantee a soul's salvation after offering absolution. Some argue that the surveyor is different because the surveyor provides work products to the client. It is true that the surveyor often provides plans, reports, descriptions, and new corner monumentation. However, these work products are no different from the abstract or estate plans resulting from the attorney's service; marriage certificate resulting from the minister's service; or medical chart resulting from the medical doctor's service. Seldom can professionals in the business of providing services offer guarantees on the outcome of their services, regardless of the number and form of the work products prepared.[7] Knowing the uncertainties in a professional practice that involve services to clients, how can the surveyor provide a guarantee on a boundary location given the lack of reliable measurements, information, and memories of local people the surveyor must call upon in order to provide his or her services? The honest answer is that the surveyor cannot and should not be expected to provide such a guarantee.

While the foregoing may seem logical and may seem to make sense, in practice, the logic is often overlooked—with members of the legal profession often being the worst abusers. Few surveyors can honestly admit that they have not been pressured by attorneys to provide guarantees on matters that for practical purposes are impossible to guarantee.

> [T]he land surveyor's work often involves retracing the footsteps of surveyors who, approximately 100 years previous, performed surveys, kept field notes and set stone monuments to establish and perpetuate section lines and corners. . . . While present-day, licensed land surveyors are required to follow local, state, and federal regulations that define present-day standards of practice while searching for ancient monuments, it is a foregone conclusion that present-day surveyors may or may not find a particular ancient monument. . . . [B]ut it is impossible for him to insure that he is, in fact, standing in the 100-year old footprints of the original surveyor. . . . It is possible for two qualified surveyors to meticulously follow the standards of practice for surveying while retracing a 100-year old survey, and disagree on a corner location without either being negligent.

6. Although the opinion is communicated in the surveyor's work products.

7. Certainly, it should be pointed out that the ceremony and cost attending a church wedding, combined with God's blessing as evidenced by the marriage certificate signed by the minister, should be a work product above any a surveyor could offer. Despite all this, no one realistically sues the minister or God for negligence in bringing couples together in marriage.

> . . . Locating and remonumenting ancient corner locations is not an exact science, and mere location of a corner, or approval or disapproval of a particular survey does not automatically establish either compliance of [sic] non-compliance with a surveyor's standard of care.[8]
>
> When a man has had a training in one of the exact sciences, where every problem within its purview is supposed to be susceptible of accurate solution, he is likely to be not a little impatient when he is told that, under some circumstances, he must recognize inaccuracies, and govern his action by facts which lead him away from the results which theoretically he ought to reach. Observation warrants us in saying that this remark may frequently be made of surveyors.[9]

Consequently, it must be recognized as a fundamental postulate that when seeking the surveyor's services, the client will receive an opinion and not an absolute guarantee. The limit of the surveyor's responsibility to a client could be stated as follows:

> The surveyor should make every attempt to communicate useful, accurate, and understandable locative information on boundaries and limits of ownership (if legally fixed); along with visible and substantial improvements located on or over the surface of the property that when coupled with the surveyor's opinion on the boundary location and information obtained from reasonably available records, would disclose matters that could or may effect the marketability of the title and that would be expected from competent surveyors, similarly situated.

### D. Surveyor's Record Search Is Different from a Title Search

A surveyor's record search is different from a title search. A comparison of the surveyor's research responsibilities with a title abstractor's responsibilities should enlighten non-surveyors. An explanation is often required because an attorney, in an effort to eliminate the surveyor's need to research the records, will provide a title abstract. An abstract is helpful but not conclusive or often complete for the surveyor's purpose. While an abstractor and surveyor frequently refer to the same documents to obtain information, the similarity ends there. The surveyor is seldom concerned with the soundness of title as evidenced by the words and sequence of documents recorded. Proper parties, acknowledgments, liens, consideration, dates, releases, and the category of the document are seldom of interest, let alone read by the surveyor. The surveyor's efforts are focused on boundary information. More times than not, the surveyor will scrutinize each

8. Yellowstone Basin Properties, Inc. v. Burgess, 843 P.2d 341, 350–351 (Mont. 1992).

9. Thomas M. Cooley, Chief Justice, Supreme Court of Michigan, 1864–1885, *The Judicial Functions of Surveyors,* 14(2) Surveying & Mapping J. 161 (1954).

word in the property description while failing to even examine the names of the parties to the deed. The title abstractor, for his part, will examine the information in a deed with great care but seldom will understand or adequately visualize the property configuration from reading the description. Furthermore, the abstractor seldom goes beyond forty years of records, while a surveyor can seldom rely on only forty years of records to obtain the information needed. Often a surveyor will search back through the records until the record creating the boundary (i.e., the parent tract) is discovered.

> And then there are those who will hand the surveyor a deed containing no information but the names of adjoining landowners, and expect him to unpack his instrument and locate their boundaries forthwith. If a starting point is in dispute or not readily found, they will know of a base line from which every valid survey should start. . . . [10]

A surveyor will often spend as much time looking at related records as he spends examining the records within the client's chain of title. Consequently, surveyors often devote considerable time seeking private survey maps, road records, maps in state archives, records in local historical societies, and original lot layouts. The abstractor seldom uses these related records. Perhaps more important, a competent surveyor will always search the records pertaining to the adjoining properties—often back to a historical time exceeding or equal to the search of the client's records. The justification is that the surveyor seeks to establish a boundary common to two or more properties—the records for one chain of title may be just as important and relevant as the records for another chain of title.

Based on this comparison, the surveyor's record search is often broader and more in-depth than the title abstractor's search. The attorney should realize that good title information does not necessarily mean good survey information and vice versa. The surveyor who is content to rely upon the client's abstract will often fail to discover problems and not properly retrace the client's boundaries.

### *E. Limitations*

A surveyor ordinarily confines the survey to those matters that would be disclosed by a reasonable search of the public records and information visible at the site or immediately adjacent to the boundaries. The surveyor must rely on normal human facilities coupled with the surveyor's specialized knowledge to discern important information. Consequently, the surveyor is not able to scan below the ground surface to see buried conduit, contaminants, or a high groundwater table. The surveyor is not capable of smelling petroleum distillates or other chemicals that were dumped long ago. The surveyor is unable to fly over large parcels or nearby surrounding property to locate wetlands, ancient graves, protected resource areas, harmful dumps, or incompatible land uses that may affect the property in

10. Richard A. Mawhinney, Surveying Problems In Eastern Maine, Coastal and Interior 8–9 (c. 195?).

question. A surveyor's hearing is incapable of detecting voltage transmissions that exceed the allowable voltages mandated in a deed. Finally, the surveyor is not a clairvoyant and is incapable of realizing what future use the property or nearby property may be subjected to. The point is that the complexities of problems that may reside on the property cannot all be identified by employing the services of a surveyor. Complete protection is only possible by the employment of numerous specialists at great expense. There is no other alternative.

In other cases, the surveyor's services will be employed under less than favorable conditions with strict time limits imposed. These limitations may further restrict the surveyor's efforts to locate or discover problems or important information. An example may be a boundary retracement in January where snow covers many of the corner markers. Another example may involve the location of property boundaries along water during a summer drought where the water level is abnormally low.

The surveyor will or should be able to identify certain problems that are appropriate to the surveyor's education, training, and knowledge and the reasonable availability and notoriety of the information—no less, and no more.

## II. Obtain a Survey or Not?

One of the questions that should come up at the time of a conveyance, logging, or erection of improvements is whether the buyer or landowner should obtain the services of a surveyor to perform a boundary retracement survey.[11] A retracement survey is a survey with the objective of locating boundaries previously established (as opposed to creating new boundaries). It is dependent upon a historical surveyor's work. The boundary defined during a retracement survey should coincide with the location of the boundary that had been previously created.

> [A] surveyor can be retained to locate on the ground a boundary line which has theretofore been established. When he does this, he 'traces the footsteps' of the 'original surveyor' in locating existing boundaries. Correctly stated, this is a "retracement" survey, not a resurvey, and in performing this function, the second and each succeeding surveyor is a "following" or "tracing" surveyor and his sole duty, function and power is to locate on the ground the boundaries corners and boundary line or lines established by the original survey; he cannot establish a new corner or new line terminal point, nor may he correct errors of the original surveyor. He must only track the footsteps of the original surveyor. The following

11. A boundary retracement survey should not be confused with a mortgage loan survey or mortgage loan inspection. Under ideal conditions, the mortgage loan inspection will identify visible and discernible encroachments at the site. At worst, the mortgage loan inspection is a drive-by view of the property to make sure a house resides on or somewhere near the property.

surveyor, rather than being the creator of the boundary line, is only its discoverer and is only that when he correctly locates it.[12]

The cost of a boundary survey can be substantial and range anywhere from five hundred dollars to several thousand dollars, depending on the property's size, terrain, vegetative cover, existing corner markers, weather, and time span since the previous survey, to name a few factors. The money for a retracement survey would be well spent if the buyer is forewarned from purchasing a serious problem. On the other hand, the money spent on surveying services can seem like a waste if no problems are discovered and the property is found to be as it appears according to the most recent records. Consequently, attorneys have often sought some aid or methodology in formulating a recommendation whether the client should obtain a new boundary retracement survey or, in the alternative, accept as sufficient a historical survey or the current record information.

## A. Evaluation Components

There are a number of components found in descriptions and plans that may be evaluated in order to arrive at an opinion whether to recommend a retracement survey be performed or, in the alternative, accept the historical survey and description. The following are the most common evaluation components.

### 1. Closure

Performing a closure check is one tool used to check for potential problems in a metes-and-bounds description. A description, if properly written, should provide enough information with sufficient precision to mathematically begin and end at the same point. A properly prepared metes-and-bounds description should have a direction and distance for each and every segment of the boundary surrounding the entire property.[13] In other words, the directions coupled with the distances should begin and end at the same point. If the measurements were plotted, they should form a closed figure. Failure to provide sufficient information for a closure check on a metes-and-bounds description is a symptom of a potential problem.

A closure check can be performed by plotting each course[14] in sequential order, as the courses are listed, to determine if the resulting figure forms a closed figure.[15] Another method is to convert the courses into dependent coordinate values[16] and compare the final coordinate values with the starting coordinate values to see if they are the same.

---

12. Rivers v. Lozeau, 539 So. 2d 1147, 1151 (Fl. Dist. Ct. App. 1989); *see also* Tyson v. Edwards, 433 So. 2d 549 (Fla. Dist. Ct. App. 1983); McKinley v. Hilliard, 454 S.W.2d 67 (Ark. 1970).

13. In some cases where the boundary follows a stream, lake, road curve, or other curvilinear surface, a "tie line," or straight line course between the termini, is provided to serve as a closure check.

14. A course is composed of a direction and distance.

15. This would only be true for metes-and-bounds descriptions and not for other forms of descriptions such as those often used to describe easements.

16. This is mostly done using coordinate geometry software known as COGO programs.

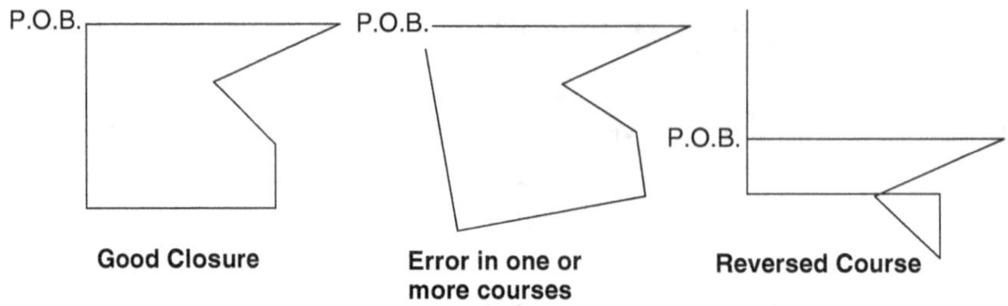


**Figure 13.2** Three figures show various results from plotting the directions and distances in a deed description. The figure on the far left is a plot that forms a closed figure, indicating an acceptable closure. The middle figure probably contains one or more errors in one or more courses. Finally, the figure on the far right represents what happens when a course is reversed (e.g., N03°W → S03°E).

The failure of the figure to close or the last coordinate values to be the same as the starting coordinate values indicates the description is either incomplete, inadequate, or contains one or more errors.[17] There may be one or more missing courses, reversed courses, transposed numbers, errors in the field measurements, or a host of other possible problems. If a problem is discovered, the historical descriptions should be compared with the present description to attempt to locate transcribing errors. If the error is in the fieldwork, only a survey can reveal the source of the problem.

Plans should not be immune from a closure analysis. There are numerous plans that show a closed figure, but when the numerical information on the plan is evaluated, the resulting figure fails to resemble the figure shown on the plan.

## 2. The Number of Exceptions

In some cases, one or more lots have been conveyed off a parent tract in a piecemeal fashion over a period of time. The present description often contains a description of the parent tract and then goes on to except one or more conveyances out of the parent tract description. The result is that the abstractor is often unsure of the extent and size of the original tract.

Example: " . . . to the point of beginning, containing 5 acres . . .

**Except** a one acre lot sold to Sarah Bunting as found in deed book 343, page 211.

**Except** a two acre lot sold to Ezekial Willoughby by deed executed 9 Oct. 1955.

Subject to a utility easements described in deed book 321, page 93. . . ."

To complicate matters, the conveyances (exceptions) out of the parent tract more times than not were made without the benefit of a perimeter survey of

17. The potential sources of the error are numerous and often frustrating to locate. Some are errors made in the field, while others are errors caused in copying or writing the description.

the parent tract. Furthermore, the subsequent surveys of the out-conveyances failed to check for conformance and continuity with the previous conveyances out of the parent tract. The result is much like an apple vendor who starts with a bushel of apples, promises apples to numerous parties, and after delivering on the promises (1) finds there are more promises than apples, (2) has no apples left for herself, or (3) discovers the apples left for herself are the worst of the bushel.

Furthermore, it must be recognized that the surveyor's fee to determine the remainder is undoubtedly going to be high because of the need to survey numerous parcels—a long overdue debt from all the previous half-measures that were adopted in the past to save money. The failure of the original owner of the parent tract to obtain and pay for a complete perimeter survey and subsequent dependent divisions will place great cost and risk on the owner of the remaining land many years later.

### 3. Number of Easements/Restrictions

Similar to the problem with numerous exceptions are the uncertainties surrounding property that contains numerous easements and restrictions. Failure to adequately locate, describe, and define the easements and restrictions may cause numerous problems. Often, easements are not adequately described and take the form of blanket easements that have no definitive location, width, or dimensions. In effect, they can fall anywhere on the property in any manner of width and orientation—blanketing the property. Utility easements are notorious for causing this problem.

> Example: "Does hereby grant to ________ utility company an easement for the installation of electric lines and poles in a manner or location as they determine."

In other cases, easements are described or shown and have some orientation and shape, but lack clear reference to the property boundaries and consequently appear to "float" somewhere on the property. Easements taken for highway purposes are notorious for this fault.

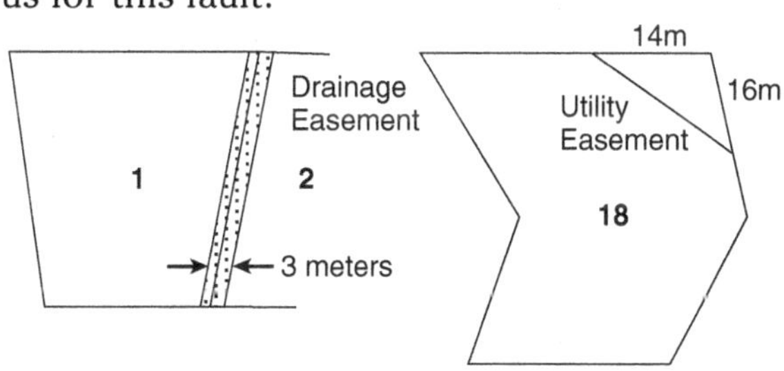


**Figure 13.3** The drainage easement between Lot 1 and Lot 2 has a width but is not located with certainty. In other words, is the easement centered 1.5 meters on each side of the boundary or some other proportion as a scale of the cross-dimension of the easement would suggest? On Lot 18, the utility easement is located but not dimensioned. What is the width of the utility easement?

A properly prepared easement should have a point of beginning located by using one or more directions and distances from a property corner. In a strip description, the location of the easement is given relative to some part of the easement using one or more directions and distances. The width of the easement is stated. Furthermore, the use or limitations of the easement are stated.

> Example: "Beginning at a 5/8th inch diameter rebar, the northwest corner of the tract described previously; thence along lot 15, South 18° 21' 22" East 45.211 meters to the center of a 4.954 meter wide drainage easement; thence along the center of the drainage easement, North 88° 14' 45" East 142.232 meters to the boundary with lot 3; lying North 18° 21' 22" West 34.571 meters from the southeast corner of the tract described, extending and contracting the ends of the easements to meet the boundary with lot 3 and lot 15. . . ."

Failure to find a complete description of the easement on the survey plan or within the description may be a prelude to serious problems.

## 4. Age of the Previous Survey

The time that has elapsed since the last survey can be relevant for evaluating the need for a new boundary retracement survey. The greater the age of the last survey, the more likely that errors or problems will be present. Even in short time periods, outconveyances can occur, improvements can be erected that cross the boundary or building setback, and incompatible uses can arise. Short of a complete title search to the original operative conveyance, items that can be used to help determine the time period of the previous survey include the following.

**Coarse Directions.** When all the bearings are given to the nearest degree or 1/4 of a degree of arc (i.e., 15°, 30°, or 45°), the survey was likely completed using a compass. The compass was the predominate survey instrument up to the early 1900s.

> Example: " . . . South 15° West. . . ."

If most or all of the bearings are given to the nearest minute of arc, the surveyor probably used a transit. The transit was the predominate survey instrument from the early 1900s to the 1960s.

> Example: " . . . North 85° 23' East. . . ."

**Coarse Distances.** Distances in rods, poles, or perch are a good indication the survey was probably performed sometime before the mid-1900s.

> Example: " . . . 23 rods . . . 23 poles . . . 23 perch. . . ."

Distances in chains and links were used in lieu of rods, poles, or perch in some cases. It was more common to use chains and links prior to 1900 than during a later time.

Example: " . . . 18 chains . . . 3 links. . . ."

**Area Not in Acres.** The use of rods, poles, or perch for the area of the parcel instead of square feet or the decimal part of an acre indicate the survey was probably performed before the mid-1900s. The use of roods to fix the area probably indicate a survey was performed prior to or during the early 1800s.

Example: " . . . 4 acres, 2 roods, 2 rods. . . ."

**Corner Monuments.** The monuments used to mark the property corners are often a good indication of the age of the survey. Presently, the preponderance of corners monumented by surveyors are in the form of concrete or metal. This was not true several decades previously. Metal was very rare and quite valuable in the 1700–1800s. As a consequence, other materials with resistance to rot and decay were often used. In rural and remote sections of the country, posts, trees, and stones continue to be used for corner markers in order to avoid carrying the weight of iron to remote sites. Without using metal, it was common to mark corners using naturally occurring material such as posts, trees, ledge, streams, stones, wood, and so forth. A survey several decades old would likely have a majority of corner monuments composed of stone or wood.

Example: " . . . to a cedar post; . . . to stones; . . . to a planted stone; . . . to a pine knot; . . . to a blazed spruce. . . ."

Urban lots that call for post corners were probably surveyed prior to 1940. Rural property descriptions that call for all non-metallic corners were probably surveyed sometime before 1970.

Certain monuments may provide a more definitive estimate of the original survey. For example, a corner marked by a chestnut indicates a survey performed sometime before 1930. The mature American Chestnut tree was virtually wiped out by the chestnut blight sometime after 1930. Other monuments, though not as common, provide similar estimates of the age of the last survey. Other monuments include coal hearth,[18] ice house,[19] buffalo waddle,[20] warpath, Indian camp, turnpike,[21] and so on.

### 5. Skill, Knowledge, and Care

Obviously, the skill, knowledge, and care of the original surveyor or the person who wrote the original description can be an important evaluation component. The greater the skill, knowledge, and care, the less chance for problems. There

18. Pits for turning trees to charcoal for use in iron furnaces (circa 1700–1850).

19. Sheds used to store ice for the summer (circa 1700–1940).

20. Pits where buffalo rolled to cake their bodies with mud and keep off insects. (circa 1600–1740 for eastern woodland bison; circa 1800–1870 for western bison).

21. Many of the earliest roads were turnpike roads privately owned and operated for profit by collecting tolls. Most states took over private turnpike roads by the 1930s.

are methods to estimate the skill, knowledge, and care of the original surveyor or at least the person preparing the description.

**No Directions.** When there are no directions other than general directions, the chances are good that a survey was not performed or the description was written by the landowner or attorney unfamiliar with or unable to make survey measurements. The use of the following terms generally indicates a lack of care or concern on the writer's part.

> Example: "easterly . . . southerly . . . northerly . . . parallel . . . perpendicular. . . ."

**Unbalanced Precision.** Sometimes a description is found where the bearings are given to the nearest degree or minute except for the last direction, which is given to the nearest second, and all distances are given to the nearest foot except for the last distance, which is given to the nearest 0.01 feet. This unbalanced precision often indicates the surveyor never checked the fieldwork and merely calculated what the proper final bearing and distance should be.

> Example: " . . . North 85° 23' East 121 feet; thence South 18° 13' 45" East 200.321 feet to the point of beginning. . . ."

Another situation where unbalanced precision of the measurements may occur is where a description was prepared by borrowing measurements for a common boundary from the descriptions of the neighboring property. The varying age of the neighboring surveys together with the varying basis for the directions often resulted in the composite description not closing as previously explained. In extreme cases, the units were not even harmonized.

> Example: " . . . N 83° E 11 rods to a stone; thence southerly 100 feet more or less to a metal pin; thence South 82° 33' 20" West 324.22 feet to a metal pin; thence. . . ."

**Omission of Measurements.** The omission of one or more measurements makes the geometry of the original survey difficult if not impossible to check. Many times, the description intentionally omitted several courses the surveyor placed on the original plan in order to reduce or shorten the tedious work in writing a description.

> " . . . thence in a straight line to the point of beginning . . . thence by the stream several courses and distances. . . ."

In other cases, the original surveyor may have intentionally omitted one or more measurements in order to prevent a closure check or verification of his work.

Finally, many descriptions have omitted one or more courses through carelessness and mistake. The common procedure employed in conveyancing is to delegate the task of preparing a deed to the lowest paid employee in the law firm. The description, often incomprehensible to the legal clerk, is copied from one deed to another without checking. This often results in mistakes, omissions, and transposed numbers going undetected.[22]

**Poor Monuments.** Sometimes the use of posts, birches, poplars, alders, wooden stakes, etc. for all the corner monuments indicate an "office" or "chamber" survey. By citing monuments that have a short span before their decay and obliteration, the surveyor has made it difficult for someone to prove the surveyor was not in the field. Even in the rare cases where such monuments were actually used or adopted, their short life span indicates a lack of care by the original surveyor and forewarns of some difficulty in retracing the boundaries.

**Distance Uncertainties.** Distances shown along with an indication of uncertainty may suggest: (1) the surveyor is not sure of the "on-the-ground" terminus location; (2) the surveyor has not actually measured the line or only measured the line using crude methods or equipment; or (3) the surveyor does not want the distance to control.

> Example: " . . . thence South 15° 18' 20" West 230 feet more or less to the shore of Moosehead Lake. . . ."

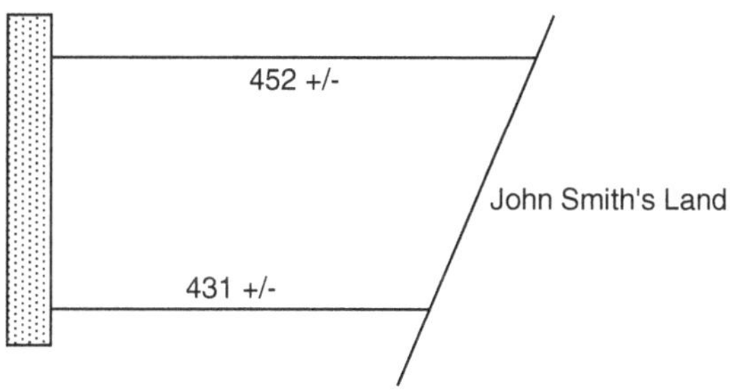


**Figure 13.4** In the figure, the distances include a +/– indicating some uncertainty in the distances.

**Low License Number.** Licensing of land surveyors in many states occurred relatively recently. In other cases, when existing licensing acts were revised to exclude engineers from performing boundary surveys, the then-current licensed engineers were allowed to apply for and receive surveying licenses.[23] Consequently, many persons were granted licenses without examination or verification of experience. These surveyors are known as "grandfathered" surveyors.[24] Of course, it must

22. A sample of ten deed descriptions by the author indicated that there was a 33 percent chance of an error in a description if the description was copied five or more times.

23. At the time this article was written, West Virginia continues to allow professional engineers to perform land surveying.

24. This term properly applies to all surveyors who were licensed under a previous edition of the licensing law regardless of the fact that they did or did not have to take an exam.

be stressed that there are numerous excellent and well-qualified surveyors who were licensed in this fashion and provided the highest quality surveying services. However, there are numerous others who were incompetent before they were licensed and remained incompetent after receiving their license. In some states, grandfathered surveyors generally have low license numbers or some letter/number designation indicating the manner in which they received their license.

### 6. Confused Obligations

Many surveyors may be competent in retracing boundaries but are not well versed in real property law. As a result, some surveyors have failed to recognize that the boundary described in the records does not necessarily coincide with the ownership boundary. In some cases, a surveyor has oversimplified the work required to locate the boundary described in the records by merely recognizing the occupation line as the property boundary. In other cases, the surveys and resulting descriptions ignored important and fundamental canons of title.[25] In still other cases, the surveyor has retraced the boundary according to the records without taking into account certain rules of construction that recognize ownership extending beyond the line of the original survey. As a result, the title may be clouded or made unmarketable, or unmarketable title may appear as marketable title. Below are common omissions.

**Public Road Easements.** The call for a public road easement generally carries title to the extent of the grantor's title or the center of the public road, whichever is less. However, the original and subsequent surveyor frequently stopped their measurements short of the center of the road and placed the boundary along the right-of-way or edge-of-pavement. If the title to the area underlying the road easement is important[26] or the building setback distances are critical, the attorney should question the scope and depth of the surveyor's research and the assumptions the surveyor used in determining the right-of-way title, location, and width. In terms of descriptions, there is likely to be uncertainty whenever the following or similar words are found in a description:

Example: " . . . to the public road; thence along the road. . . ."

In defense of surveyors who do not clarify the location of their client's ownership along public road easements, it should be recognized that road records are more times than not located in places other than the registry of deeds, such as at municipal offices. In these offices, the records are often placed in some out-of-the-way location, such as the attic, basement, or broom closet. In many cases, the road records have been misplaced, lost, or destroyed. When the records have been located, they are often difficult to use because they lack

25. A "stranger to the deed" problem is one example where the person preparing a description may have ignored an important canon of title.

26. *See, e.g.*, Lamb v. Euclid Ambler Assocs., 563 A.2d 365 (Me. 1989); Metzenbaum v. City of Carmel-By-The-Sea, 44 Cal. Rptr. 75 (Cal. Ct. App. 1965).

indices or are based entirely on long-lost features or long-forgotten names. The road descriptions themselves are ambiguous and difficult to read—written at a time long before typewriters, when the land was mostly farms and all the people were familiar with everyone in the county. Research is tedious, time consuming, expensive, and often unproductive.

> Example: "Beginning along the Enderson Turnpike, 20 rods south along the pike from William Jackson's hog pen; thence S82°E 20 rods to a post; thence . . . to end at the lane leading to Ezekial Thompson's Grist Mill. The road to be opened at a width of 2 rods."

Other common assumptions made by surveyors that may cause problems along roads are the following:

a. The surveyor may have fixed the width of the road to be the average width between improvements or assumed a statutory or common law width for the public road easement or right-of-way (e.g., 3 rods). The surveyor has in most cases *not* researched the public records to determine the legal width. (This is not the case where the surveyor has shown a book and page number where the road record can be found.)
b. The surveyor has probably assumed that the title stops at the right-of-way rather than extends to the center.
c. The surveyor has probably assumed the center of the traveled way is the center of the easement.

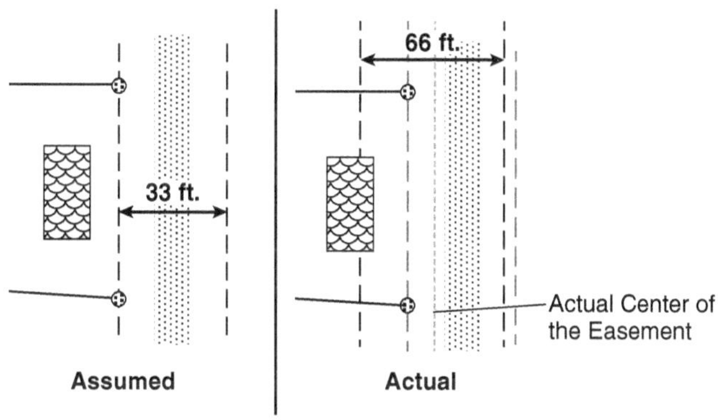


**Figure 13.5** The figure illustrates the common assumptions that are made and what the actual records, when located, reveal. The figure at the left shows how the surveyor has assumed the asphalt roadway is in the center of the easement and the easement is 2 rods wide (by common law for undefined roads). The figure on the right is the results after the record reveals the asphalt roadway was placed inside the easement but not in the center. Furthermore, the records show the easement is 4 rods wide. The end result is that the house resides within the easement.

**Water Bodies.** Unable or unwilling to wade into the water, the surveyor most likely made his measurements along the water's edge. In depicting or describing the boundary, the surveyor has shown the ownership boundary to coincide with the survey line. While there is considerable dissension among courts, the inadvertent call for the bank or shore in a description does not always convey title to the center or low-water mark. The attorney should be cautious when the following or similar words are found in a description or on the plan:

Example: " . . . along the bank . . . shore . . . edge. . . ."

As a result of past loose practices in surveying and writing descriptions for riparian property, valuable shore or riparian property has lost some of its value because the owner cannot enjoy the full extent of water recreation or access. The following are some situations that are ripe for problems:

a. Great Ponds or Navigable Lakes—Many surveyors either stop at the existing water's edge or assume the title stops at the high-water line or top of the bank.
b. Non-tidal Streams, Rivers, etc.—The surveyor shows the boundary along the water's edge or bank when in fact it should be in the center[27] or along the low-water line.[28]
c. Tidal Water—The surveyor shows the boundary along the shore or upland edge of the flats when in fact the boundary should have been along the low tide line.[29]
d. Side Boundaries—Even when the boundary is shown to coincide with the extent of ownership, the issue of side boundaries is often overlooked or ignored. The attorney should question how the boundary is to be or was extended between the limits of the survey to the limits of ownership.

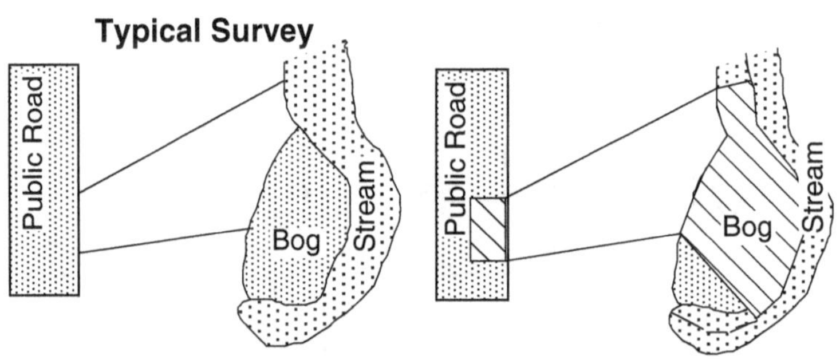


**Figure 13.6** The figures show the sometimes considerable area that is omitted from the survey and subsequent description prepared for property bordering water bodies and roads.

27. Some states hold the center of the stream is a line equidistant between the low-water's edge. Other states hold the center to be a line that follows the deepest point in the channel.

28. The location of the limits of ownership varies between states. Most states recognize that private title extends to the low-water mark on navigable bodies of water and the center for non-navigable bodies of water.

29. The location of the boundary may depend on state law. For example, in Maine, title extends to the low-water line or 1,650 feet from the high-tide line (whichever is less).

e. Common Law Easement—Often the surveyor fails to show the common law easement that may exist between the high and low water. Even in cases where the landowner's title extends to the center or low-water line of a stream, there may be a common law easement in favor of the public along the water body.

## B. Incompatible Standards

In addition to assessing the quality of the original description or survey, persons relying on the survey or description should understand that the standards of the former surveyors and their surveys may not be compatible with the needs of the current landowner or prospective user. For example, the last survey performed when the property was farmland at $2 an acre is seldom adequate to rely upon when building a $3 million mall.

### 1. Different Types of Services

Surveyors do more than simply locate boundaries. Surveyors are able to and often provide a wide assortment of different services.[30] Surveying services often include construction surveying, control surveys, development planning, land information system management, mapping, and numerous other services. Often, these services take advantage of the surveyor's knowledge of boundaries and measurements yet fall far short of or go beyond the surveyor reestablishing the boundary location. Two common services that surveyors provide that use boundary knowledge but are not considered boundary retracement surveys are mortgage loan inspections and ALTA/ACSM Land Title Surveys.

The necessity of removing the survey exception from a mortgagee's title insurance policy has led to a service called a mortgage survey, mortgage loan survey, or mortgage loan inspection. This service is not a boundary survey and is inadequate and inappropriate for erecting improvements, discovering title defects (other than obvious encroachments), or preparing a revised description. Some states have adopted standards for these services; most states have none and rely on the surveyor and client to reach agreement on the standards to apply. A few states have determined this service to be so marginal and the resulting product so defective or misleading that the services should be discouraged, upgraded, or outlawed. Even states that have adopted minimum standards recognize the services provided fall far below what is reasonable and common for a boundary retracement survey. Many prudent surveyors will clearly note the limitations of this service on the face of the sketch or report that is prepared. Despite the notes, the sketches are frequently confused with boundary retracement surveys and used where only boundary retracement surveys should be used. Unfortunately, the form is often confused for substance. The use of the words "sketch," "mortgage,"

30. This has led some universities and colleges to change the name of their surveying programs to names such as "geomatic engineering," "spatial land information science," etc.

or "inspection" in the title or name is usually a good indication that the document does not represent a boundary retracement survey and should not be relied upon.

An ALTA/ACSM Land Title Survey generally involves a survey of the boundary, but a boundary retracement survey is not an ALTA/ACSM Land Title Survey. An ALTA/ACSM Land Title Survey is ordinarily sought to remove the survey exception from title insurance policies for commercial property, property valued over $1 million dollars, or property where the homeowner seeks a title insurance policy without the survey exception. Consequently, the ALTA/ACSM Land Title Survey goes far beyond showing the location of the boundaries. The specifications that apply to this type of service require the surveyor to note conditions, improvements, and objects that may affect the marketability of title. The fee for performing the scope of services required for this level of service generally exceeds the cost of a typical boundary retracement survey. The attorney who expects this level of service should not seek a boundary retracement survey.

## 2. Different Classes (Levels) of Boundary Surveys

There are different classes of boundary surveys. For laypersons unfamiliar with professional services, there is a belief that all professional services are uniform and consistent regardless of the client, circumstances, or property involved. The layperson's belief is true in part, because a certain minimum level of service is always provided. Beyond a minimum level, however, the surveyor's services vary considerably, depending on situations and conditions surrounding the services. An analogy can be made to the attorney preparing an estate plan. All clients seeking an estate plan can expect the attorney to prepare a will. However, a client with $6 million worth of assets will receive a markedly different level of service than a client with $40,000 worth of assets who also seeks an estate plan from the same attorney. The resulting work products may both include a will, but the wills are likely to be markedly different in their content, number of pages, and scope.

In boundary surveys, the level of service will often differ depending on the situation, location of the property, and needs of the client. The landowner who seeks a boundary retracement survey prior to logging will likely receive a different level of service than a client who wants a boundary retracement survey in contemplation of litigation or subdivision. Another way to comprehend the different levels of a surveying service would be to compare a boundary retracement service for a 0.25-acre city lot where a $1.5 million building is being erected and the boundary retracement service for a 400-acre parcel of woodland in the Appalachian mountains where a new hunting camp is being erected. It follows that the boundary retracement plan or description prepared for the woodland owner should not be used at some later time to subdivide the property or erect valuable improvements (e.g., a new Walmart).

The layperson should be told that regardless of the situation, the location of the property, and the needs of the client, the surveyor has probably completed adequate research and made a reasonable attempt to locate the position of the corners. The level of surveying service frequently varies regarding (1) the refinement and precision of the retracement measurements; (2) the extent of verification with other corner monuments and evidence found on adjoining parcels; (3) records examined outside the registry of deeds; (4) markings made along the boundary between the corners; (5) the location and depiction of the size, visibility, and type of encroachments or appurtenances affecting the property; (6) peripheral services such as locating easements, floodplains, setback lines, and the like; and (7) the quality, number, and scope of the document(s) prepared that are meant to communicate the surveyor's opinion.

Standards promulgated by licensing boards and professional societies often make a point of classifying the different types and levels of surveying services (or permit the client and surveyor to do so). It would be prudent for any real estate practitioner to obtain a copy of the current local and national surveying standards to better understand the differences in services when evaluating or ordering surveying services.

### 3. Evolving Standards for Boundary Surveys

Over time, the minimum standards or criteria for boundary surveys have changed. In most cases, the minimum standards or criteria have toughened, resulting in higher standards and better quality work products than in the past.[31] Better education, training, equipment, and interaction among practitioners have allowed better services and an ever-expanding expectation about the surveyor's services from clients and users. Generally, criteria change or have been modified so that the quality, scope, and services provided several years ago are no longer the same for the present time. Consequently, the level, scope, and reliability of previous surveying services performed by the same surveyor can be expected to change with time. This situation often leads to distrust of or dissatisfaction with surveyors by title attorneys who seek to have relatively recent surveying documents recertified by the author of the original—only to have the author-surveyor demand additional time and a seemingly large fee for what was assumed to be merely an administrative matter. In the surveyor's defense, a competent and ethical surveyor will have to perform some minimal level of record research and fieldwork to update his earlier work. Similar behavior is expected from all professions. The lawyer cannot be expected to recertify an earlier abstract he completed without updating a record search. A doctor will not use a historical X-ray to certify a patient's current health for insurance purposes. A minister will not allow a historical confession to permit absolution at the present time.

31. The same attorney that conscientiously revises a client's estate plans every five years to take advantage of revised tax laws will be incensed at the surveyor for refusing to recertify a seven-year-old survey without reperforming some of the record search and fieldwork.

## III. Evaluating Title Using Survey Documents

As a general rule, three documents are prepared from a boundary retracement survey. These documents are (1) the plan, (2) the description, and (3) the report. The last two are frequently omitted in some parts of the country, depending on the circumstances surrounding the boundary retracement or the agreement with the client.

The plan, also known as a plat, draught, or survey, is a summary of the practitioner's findings or opinion. Too much detail and the plan becomes incomprehensible. Too little information and the plan loses credibility and fails to convey critical information for foreseeable reliant parties using the plan in a reasonable manner. Because the plan may be recorded, sensitive information, information harmful to the client, or information that has little or no public meaning is ordinarily not PLACED ON THE PLAN, BUT IS INCLUDED in the report.

The report, known as a "survey report" or "forensic report," varies in its content, scope, depth, and format, depending on the surveyor and the demands made upon the surveyor. The comprehensive report will contain a detailed explanation, analysis, and factual basis for the decision. Reports, unlike plans, will seldom be recorded, so sensitive, harmful, or technical information is often placed in the report. It should be noted that most attorneys fail to ask for or seek a report from the surveyor and consequently omit a valuable tool for evaluating the surveyor's work and discovering potential title problems.

Finally, the description is a written narrative of the practitioner's opinion. The description is a narrative summary of the surveyor's opinion on the record boundary location. The writing skills and consequently the quality of the descriptions vary widely between surveyors and across the country.

### *A. Plan Notes*

Often potential title problems discovered by the surveyor are disclosed on the face of the plan. The surveyor will show encroachments that cross the boundary, fences on or near the boundary, utilities that cross the property, joint driveways, and so on. Typically, these problems will be noted in pictorial form on the plan.

In other cases, the surveyor will put notes on the plan that allude to a problem or where other important information, including problems, can be found.

Notes:

1. Wetlands are present on the property.
2. No attempt was made to determine the width of the public road easement.
3. For important and relevant information, see the Survey Report.
4. There is a fence two feet south of and along the entire length of the northerly boundary.

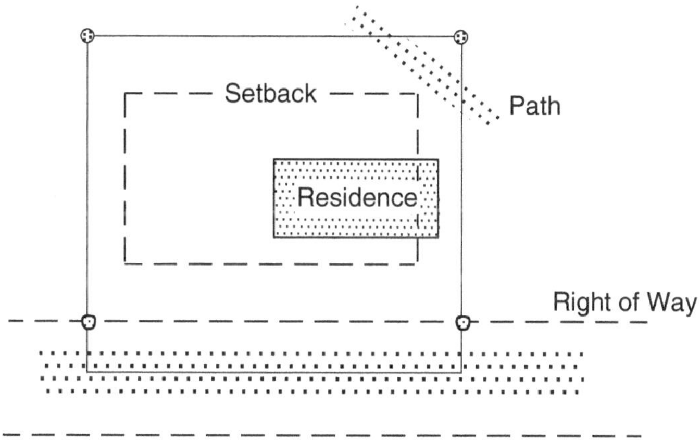


**Figure 13.7** In the figure, the surveyor has shown a path crossing the property and a residence astride the building setback line. Arguably, the surveyor may be relying too much on the intuition and knowledge of the client or attorney in communicating the message that the pictorial depiction represents a problem.

The attorney should examine the notes carefully. Often, the notes reveal problems.[32] In some cases, the absence of notes should be cause for concern. For example, the failure of the surveyor to mention wetlands or the 100-year floodplain limit for property along a stream where wetlands and flooding is expected should raise questions and cause further investigation on the attorney's part.

## B. Application of Legal Principles with Site Conditions

Aside from obvious problems noted on the plan or plan notes, an attorney should be prepared to review the plan or report in conjunction with a title abstract, apply his knowledge of real property law, detect potential problem areas, and provide recommendations or advice. Some common problems that are not noted by the surveyor but may be determined from the information shown on the surveyor's plan include the following.

### 1. Easements by Necessity

The surveyor's research will often encompass all or part of the surrounding properties. In some cases, problems with the neighboring property may point to a title defect on the client's property. The surveyor performing a survey of the grantor's remaining land may show the original lot configuration or parent tract configuration. When surrounding properties are shown, the attorney should examine the

32. At least two surveyors have told the surveyor that problems shown on plans often are ignored. However, when surveyors have color-highlighted the problems on the plans before giving them to the client, the queries have doubled, indicating many notes or problems are overlooked or not carefully examined.

documentation carefully to discover potential easements by necessity or implied easements.

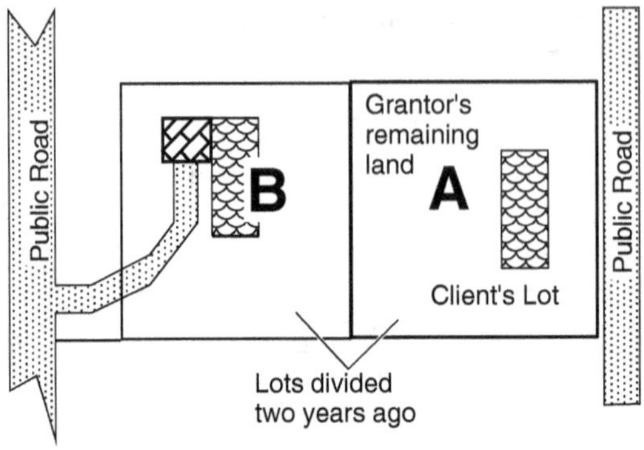


**Figure 13.8** The figure shows the existing driveway for Parcel B extending across the lands of another person in order to reach the public road—an actionable trespass. Consequently, the attorney should realize that any legal access for Lot B must pass through the grantor's remaining land (Parcel A)—indicating an easement by necessity. This condition would likely only be revealed by an astute attorney examining a detailed surveyor's plan prepared as a result of an extensive survey by the surveyor, coupled with the realization that the driveway is an actionable trespass. A title search confined to the grantor's remaining land (Parcel A) would not likely reveal the easement by necessity across Parcel A.

Consequently, the attorney should seek to have the surveyor reveal as much information about the adjoining parcels as possible and include any problems that surface regarding the neighbor's property. Extensive survey work is required to disclose the basis for this type of problem and is therefore likely to be costly for the client.

## 2. Implied Easements

In many cases, implied easements may exist and may represent an encumbrance or restriction on the reasonable enjoyment of the property. The attorney must be especially diligent for implied easements where (1) the surveyor shows or the deed describes the parcel abutting a private road, (2) the lot is shown on a subdivision plan displaying one or more streets, and (3) the surveyor shows one or more apparent encroachments (e.g., a septic field) that may have existed at the time of the original division. Often the surveyor will not mention the possibility of implied easements or state with certainty that a trespass is occurring. In fact, the surveyor may inadvertently create implied easements.[33] The following are some common examples of implied easements that may be revealed by examining the surveyor's work products.

33. The reason for this is that the typical surveyor is not trained to give title opinions and recognize title problems.

### 3. Call for Grantor's Private Road

Easements may be implied where the description calls for the grantor's private road[34] or the subdivision plan shows the grantor's private road as a boundary.

> Example: " . . . to the lane leading to the grantor's residence; thence along the grantor's residence. . . ."

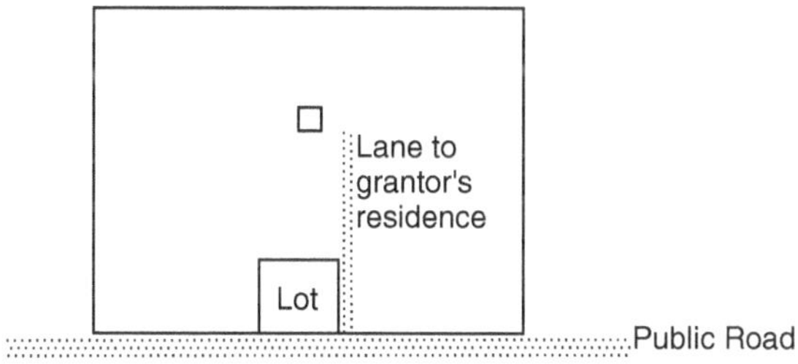


**Figure 13.9** In the example and figure, the lot is described or shown bordering the grantor's private lane. Under the common law, the lot obtains an implied easement to use the lane—necessity is not a factor.

**Roads on a Plan of Subdivision.** Easements may be implied where a buyer purchases a lot according to a plan of subdivision showing roads that were not public roads at the time of the original conveyance from the developer. Most courts in dealing with this situation have found that the lot owner has an implied easement in all the roads shown on the subdivision plan regardless of their actual location or usefulness to the lot in question.[35] This implied easement will exist whether the roads are subsequently accepted as public roads or later vacated or discontinued. There are numerous former public roads that were vacated or discontinued, and private easements remain to benefit the owners of the subdivision lots. Many of these former public roads now have expensive improvements erected upon them.

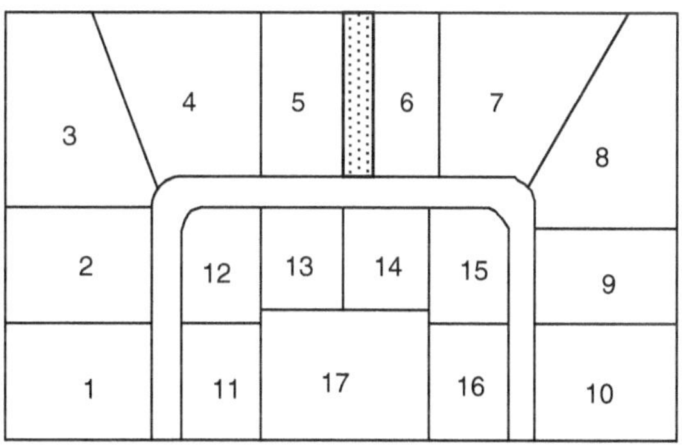


**Figure 13.10** In the figure, Lot 17 may have an implied easement in the former public road between Lots 5 and 6 (shaded). Consequently, any improvements erected by the owners of Lots 5 and 6 that intrude into the easement may be an actionable trespass in favor of Lot 17 and other lots.

34. *See, e.g.*, Robertson v. Robertson, 197 S.E.2d 183 (Va. 1973).

35. *See, e.g.* Leininger v. Trapizona, 645 A.2d 437 (Pa. Commw. Ct. 1994); Callahan v. Ganneston Park Dev. Corp., 245 A.2d 274 (Me. 1968).

The attorney should be extremely cautious when a survey plan shows that a lot owner has erected improvements on a vacated or discontinued public road that was once part of a subdivision. Attorneys should be aware that seldom does a surveyor show encroachments or adverse use of an easement that may be appurtenant to a subdivision lot where the road and lot are not adjacent to each other. In other words, the surveyor is not apt to drive around the subdivision to examine all the original easements looking for encroachments or adverse uses.

**Visible Easements.** A properly prepared plan should disclose areas of potential adverse use that are visible on the site or of such notoriety that they may be an implied easement at the time of the original division and conveyance.[36] Common easements that may fall into this category include party walls and joint driveways. The title attorney should carefully investigate the historical nature of all improvements shown on the surveyor's plan that are erected on or astride the common boundary and otherwise appear as encroachments.

### 4. Common Law Easement

Along navigable bodies of water there is generally a public easement that allows limited use by the public.[37] In other cases, where the body of water is not navigable or created by private parties (e.g., a hydroelectric dam), trespass may be prevented, even upon the water itself. Consequently, when the survey plan shows the property bordering or astride a water body, the title attorney should seek to determine whether there is a public easement or recreation may be restricted to the boundaries of the parcel in question.

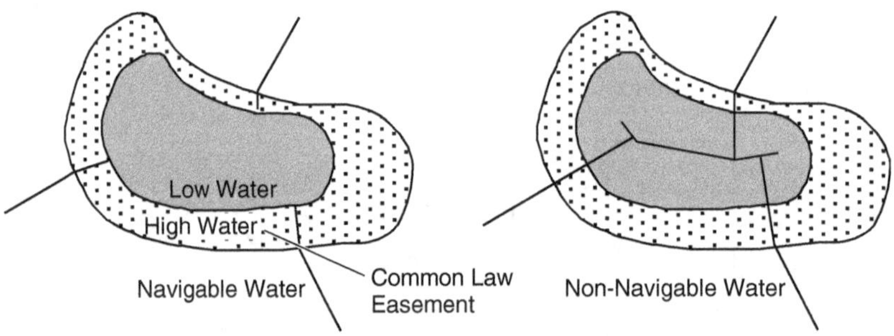


**Figure 13.11** The figure illustrates some of the differences encountered between water bodies that are navigable and non-navigable. In navigable water bodies, the public has a right to use the entire area between high-water boundaries. On non-navigable water, persons other than the fee owner may not cross the boundary without risking trespass.

There are other common law easements, though not as notorious. In certain jurisdictions there are common law easements (or licenses) to remove nuisances,

36. *See, e.g.*, Koehler v. Price, 562 N.E.2d 370 (Ill. App. Ct. 1990).

37. The major use is the right of passage but may include fishing, hunting, and certain forms of recreation.

trim boundary trees (ordinarily boundary trees are considered to be owned as a tenancy in common) or party walls, cross intervening land to reach great ponds,[38] and so on. Often, the only evidence of the use of common law easements are paths, physical marks, or improvements disclosed on survey documents.

### 5. Width of Easements

Many easements have been created without noting a width. This is often the case when prescriptive easements or utility easements such as electrical, gas, water, and sewage are present.[39] Whenever the surveyor shows utilities crossing or near the property boundary, the title attorney should determine what the width of the easement is and what limitations are in place within the easement area. In at least one case, the location of a utility line on the adjacent property prevented development of the neighboring property.[40] Whenever a utility is shown crossing the property or a deed allows a "blanket"[41] easement, the attorney should be sure the utility improvements have been properly located and the express or reasonable width for the easement will not adversely affect the fee owner's enjoyment of the property. In some cases, the landowner may need to be cautioned not to erect improvements in the easement area or advised to remove existing improvements out of the easement area.

### 6. Former Public Roads

Because the records for public roads are often located in municipal offices and inadequately indexed, road records are not often examined in a title search. The only evidence of former public roads is ancient road beds or trails shown on a survey plan. Whenever the surveyor shows the location or evidence of a former public road, the attorney should scrutinize the facts carefully. In some cases, private easements remain in discontinued or vacated public roads. Private easements may remain in public roads because they existed prior to the road becoming public. In other cases, the private easements arose by operation of law after the abandonment or discontinuance of the public road. There are also numerous cases where a public road was improperly discontinued[42] and must be presumed to continue as a public road. Another problem that may arise is where utilities have been installed in public easements. Subsequent discontinuance or vacation of the public road may cause the utilities to become a trespass upon the fee owner.[43] Records and survey maps should always be examined for evidence of former public roads and any prior or subsequent use of the former public roads.

38. *See, e.g.*, ME. REV. STAT. ANN. tit. 17, § 3860.

39. *See, e.g.*, Zettlemoyer v. Transcontinental Gas Pipeline Corp., 617 A.2d 51 (Pa. Commw. Ct. 1992).

40. Cunningham v. Otero County Elec. Cooperative, Inc., 845 P.2d 833 (N.M. Ct. App. 1992).

41. The term "blanket" easement applies to easements that can fall anywhere on the property until such time as they are physically located by construction of the utility.

42. Public roads may be improperly discontinued as a result of failure to follow statutory procedures, inadequate notice, and so forth.

43. *See, e.g.*, David L. Baird, 388 A.2d 313 (Pa. 1978); *In re* City of Altoona, 388 A.2d 313 (1978).

### 7. Encroachments in Easements

Oftentimes, encroachments in the easement are overlooked by the surveyor and title attorney. Whenever an easement is shown or described, the use must be determined from the wording of the grant or what would be reasonable under the circumstances. Any limitation or restriction not allowed by the grant, license, or law should be noted.[44] The attorney should carefully examine the terms of the deed or dedication with the information shown on the survey plan. A careful examination will often indicate unlawful limitations, restrictions, or trespass on the easements. Encroachments may take the form of (1) the servient estate encroaching upon the dominant estate by preventing or restricting the reasonable or necessary use of the easement; (2) a third party encroaching upon either the servient estate, dominant estate, or both; or (3) one dominant tenant infringing on the rights of other dominant tenants.[45]

### 8. Overburdening

When the dominant estate uses the easement beyond the express conditions, a reasonable manner, or its intended use, the easement is said to be overburdened.[46] To determine if overburdening has occurred, the attorney must not only be aware of what the original grant allows but also should check on the current use of the easement as revealed by the survey plan or report. Some common forms of overburdening that may be revealed by an inspection of survey documents coupled with a title abstract include the following.

**Utilities in Access Easements.** Often it is presumed that a person who has an access easement to reach their property (i.e., an appurtenant easement) has the right to install utilities within the easement. This use is not always allowed. Whenever property is accessed by a private easement, the attorney should check the language in the operative conveyance and the state law to determine whether utilities are allowed in the access easement.[47]

---

44. *See, e.g.*, Akers v. Canas, 601 So. 2d 305 (Fla. Dist. Ct. App. 1992).
45. *See, e.g.*, Poire v. Manchester, 506 A.2d 1160 (Me. 1986).
46. *See, e.g.*, Avery Dev. Corp. v. Village by the Sea Condominium Apartments, Inc., 567 So. 2d 447 (Fla. Dist. Ct. App. 1990).
47. *But cf.* Pope v. Muth, 481 A.2d 355 (Pa. Super. Ct. 1984).

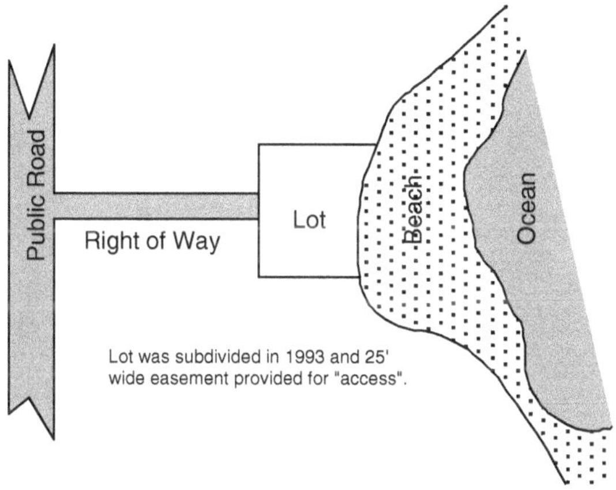


**Figure 13.12** In Maine, utilities are not included by implication in access easements created or granted after 1989.[48] Consequently, the above lot may not be able to use the access easement to bring in utilities found along the public road.

**Non-Appurtenant Parcels.** An appurtenant easement may only be used for the benefit of the appurtenant parcel. This may seem like a straightforward application of a long-tested common law rule. Nevertheless, there are numerous examples where the appurtenant easement is used to access non-appurtenant parcels by the owner of the appurtenant parcel—effectively overburdening the easement. The most common example is where an owner of an appurtenant parcel purchases additional surrounding land contiguous to the appurtenant parcel and uses the appurtenant easement to reach the additional purchases. The use of an appurtenant easement to access non-appurtenant land causes a trespass on the servient estate.[49] (To further complicate matters, when this situation is revealed, the attorney should recognize that an easement by necessity may exist over the lands of the neighbor to access the non-appurtenant land.)

In 1903, Parcel A was subdivided out of Parcel D. An appurtenant access easement was provided for the use of Parcel D. In 1960, the owner of Parcel A bought some additional land from the owner of Parcel C (shown as Parcel B). Current tax maps show Parcel A and Parcel B as one lot.

48. *See* ME. REV. STAT. ANN. tit. 33, § 458.

49. *See, e.g.*, Cooper v. Sawyer, 405 P.2d 394 (Haw. 1965); Kanefsky v. Dratch Const. Co., 101 A.2d 923 (Pa. 1954).

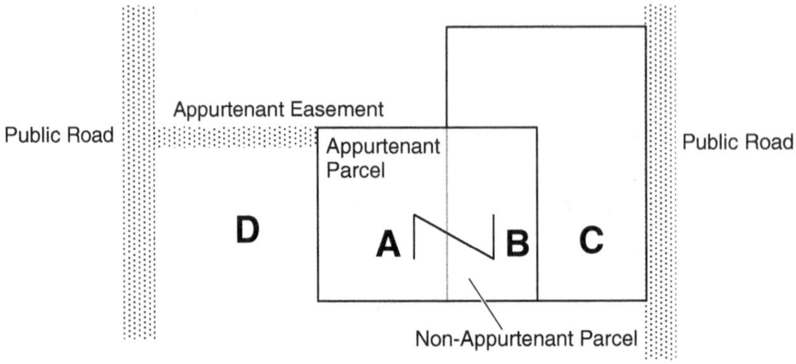


**Figure 13.13** The common owner of Parcel A and Parcel B cannot use the easement appurtenant to Parcel A in order to access Parcel B. Consequently, a somewhat ludicrous result ensues where the common owner, rather than crossing the common boundary between Parcel A and Parcel B in order to reach Parcel B, must travel by public road around to Parcel C and drive across Parcel C (easement by necessity) in order to reach Parcel B.

**Prescriptive Easements.** Often the surveyor will show an undocumented and apparently adverse use upon the client's property. Long continued adverse use, maintained under certain conditions, may give rise to an easement known as a prescriptive easement.[50] Prescriptive easements are construed strictly against the adverse user, her heirs, and assigns. Overburdening of the prescriptive easement may occur whenever the easement is used in a manner inconsistent with the use placed upon the easement during the prescriptive period. The use during the prescriptive period establishes the allowable use of the prescriptive easement thereafter.[51] For example, if an easement is acquired by prescription over a period of years by a farmer who traveled across the land of another to remove crops and repair fences, the resulting prescriptive easement may not ordinarily be expanded at some later time to provide access to residences that are located upon the former fields.

## 9. Restrictive Uses

The existence of certain natural features or improvements on or near the property frequently imposes restrictions on the property. The restrictions would not be revealed by an extensive title search. They would only be revealed by an extensive survey that shows natural features and improvements residing on the neighbor's property on or near the boundary. In other cases, restrictions are imposed through covenants. The following are common examples of restrictions.

---

50. *See, e.g.,* Comber v. Inhabitants of the Plantation of Dennistown, 398 A.2d 376 (Me. 1979); Robertson v. Robertson, 197 S.E.2d 183 (Va. 1973).

51. *See, e.g.,* Crane v. Hayes, 417 S.E.2d 117 (W. Va. 1992); Cunningham v. Otero County Elec. Cooperative, Inc., 845 P.2d 833 (N.M. Ct. App.1992); Mahoney v. Devonshire, Inc., 587 A.2d 1146 (Md. Ct. Spec. App. 1991).

**Well/Septic Field Setback.** A careful boundary survey should reveal improvements on the neighboring property immediately along or near the common boundary. In some cases, the type and scope of the improvements may work as a restriction on the neighboring property, severely limiting the location of contemplated improvements. In a few cases, it is conceivable that a vacant lot may not be improved in a reasonable manner as a result of the improvements erected on the neighboring lots. Two common improvements that often work to the disadvantage of a neighboring lot are a well or septic field. As a result of health concerns, wells and on-site septic systems must adhere to a mandatory separation distance. Consequently, if the neighbors locate their well or septic system too close to the common boundary, the neighboring landowner may be prevented from subsequently locating their well or septic system in an advantageous location.

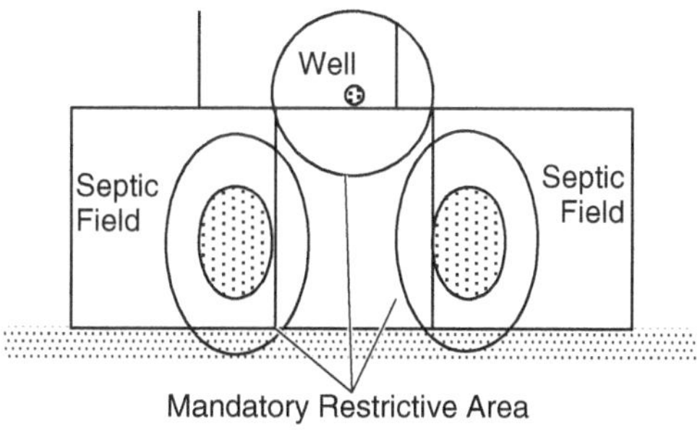


**Figure 13.14** In the figure, the neighboring septic fields and well have restricted the middle lot to such an extent that there will be little discretion allowed in the location and erection of improvements on the middle lot.

For small vacant lots where the buyer intends to erect a residence using on-site water and sewage systems, the location of neighboring septic fields and wells should be noted in order to discover burdensome restrictions or limiting factors.

**Zoning.** Records at the courthouse seldom reveal zoning restrictions. Zoning is generally a matter handled at the local municipal level. A prudent attorney will check the zoning on the property before recommending its purchase or sale. The attorney should seek the answer to one or more of the following questions regarding zoning: (1) Will the zoning allow for the contemplated improvements or use of the property? (2) Are there existing nonconforming improvements or use of the property? (3) If there are existing nonconforming improvements or use of the property, are the improvements or use allowed by variance or "grandfathered?" (3) Are there nonconforming improvements or use on the neighboring property such that the value of the property in question is affected? (4) If there are nonconforming improvements or use planned or presently existing that are not permitted, what is the cost of remediation (removing the nonconforming use or seeking a change in the zoning)?

The aid of a surveyor is enlisted to show the type and location of existing improvements or the location of particular zoning restriction lines (e.g., building setback). The surveyor can be particularly valuable in locating zoning that is site-dependent.[52] Zoning that is commonly site-dependent is shoreland zoning or resource protection zoning. Many federal, state, and local municipalities have passed legislation or ordinances designed to protect sensitive habitats such as streams, lakes, rivers, wetlands, aquifers, wildlife nesting areas, endangered species, and so on. This zoning is particularly troublesome to the title attorney because the restricted areas are not always mapped and are generally dependent on site factors such as topography, plants, resident wildlife, and geology, to name a few. In other words, there is seldom constructive notice, and reliance is almost completely dependent on actual notice. Actual notice requires the discovery and identification of the protected or sensitive habitat itself. Furthermore, mandatory buffer strips around these habitats are often large. The sensitive habitat may not be found on the neighboring parcel yet may encumber the use of the client's parcel.

**Covenants and Conditions.** Many subdivision plans contain notes that act as covenants or otherwise lead to important items that may affect the marketability of the title. Ordinarily, the wording that comprises a covenant does not raise a title concern. Seldom do the notes or covenants found in deeds or on plans, by themselves, cloud the title. However, covenants and notes, coupled with the location, size, type, or manner of improvements, often jeopardize the title. The attorney must identify covenants and compare the use shown on the survey plan with the covenants imposed on the property.

1. Lots shall be used for SEASONAL Single Family Dwellings only, NOT including Mobile Homes.[53]
2. Lots shall Not be further divided.
3. The Town . . . shall Not be responsible for the maintenance of any Road shown on this plan.
4. Lot owners shall be provided deeded access to lots over land of ________ and/or land of ________ along the 50 foot Right of Way as shown.

The above example shows poor wording and how much covenants cover.

## *C. De Minimus Encroachments*

After describing how to best recognize and use what is shown by the surveyor, the attorney should be made aware that there are often items that are not shown by the surveyor. There are numerous instances where small encroachments exist that are not noted on the survey plan. Even when they are revealed by the surveyor, they are often ignored by title attorneys. However, there will be circumstances where even small encroachments may jeopardize the marketability of the

52. The normal extent of the surveyor's services do not ordinarily include the identification of restrictions caused by natural features or habitat.

53. This is an actual note found on a plan. The manner in which the note was written prevents the year-round occupation of a $200,000 house while allowing the year-round occupation of a $15,000 mobile home.

title. The following are typical minor encroachments that are often ignored by surveyors.

### 1. Building Footprint versus Dripline

The vertical fronting of an improvement is seldom flat throughout its vertical length. The protrusions from the surface or slant of the facing may introduce minor encroachments that are not taken into account. A common example is the roof overhang (dripline) or footers (footprint) along a building's face. Where the wall of the building was constructed close to the boundary or building setback line, these protrusions out from the wall may encroach into the neighboring property or restriction line.

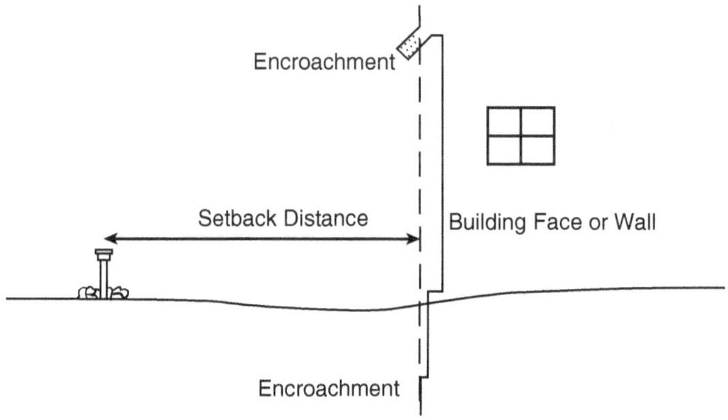


**Figure 13.15** In the figure, the eaves and footer both protrude beyond the building setback, representing an actionable violation. However, it must be recognized that even if a code enforcement officer is made aware of the encroachments, few would prosecute such minor violations with vigor.

In some cases, surveyors will fail to show porches, small outbuildings, or minor improvements that are semi-attached and extend out from the main structure. In other cases, improvements are often ignored if the improvement can be easily moved (e.g., doghouse, "handyman" shed, swingset, rosebush, etc.). Nevertheless, any improvement may be an actionable encroachment. The title attorney should be cautious in discounting even minor encroachments where circumstances suggest that even small encroachments may lead to, EXACERBATE, or escalate disputes or official citations.

### 2. Utility Lines to the House

Another common encroachment that is frequently overlooked by surveyors are overhead[54] utility lines leading to a building. In some cases, the surveyor fails to look overhead and will miss aerial wires. In other cases, aerial wires are so common in areas that to make note of them would appear ludicrous.

54. Underground utility lines are always overlooked since they cannot be seen.

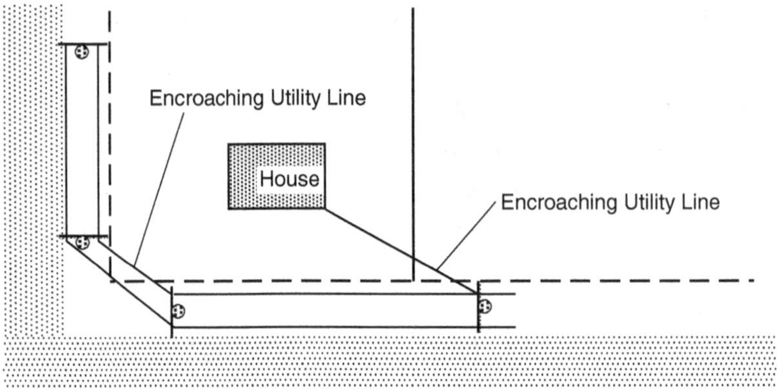


**Figure 13.16** The figure shows two common aerial encroachments that can occur. One is the aboveground wire connection from a nearby pole to the building, crossing outside the easement boundary across neighboring property. The second is the wire(s) that crosses the corner of the property, between poles located within the right-of-way.

These apparently insignificant encroachments, to neighbors quarreling or the disgruntled landowner looking for a righteous cause to obtain retribution for other slights, may be cause for legal action and subsequent removal. Many times, such minor encroachments can be easily rectified with a license that is obtained without cost or fanfare at the time of the sale or purchase of the property.

### 3. Private Use in a Public Easement

Ordinarily, continuous occupation or permanent use by a private party maintained in a public easement is an actionable trespass. This situation is not uncommon but is often difficult to detect. Furthermore, the problem may be compounded by the appearance of a license from public authorities. A common example is cable television. Congress has passed legislation[55] allowing cable television the supposed right to "piggyback"[56] on other utilities to provide services to subscribers. In some districts, this legislation has been found to be limited[57] while others have found this to be allowable.[58] Nevertheless, the cable television companies appear to have taken the position of occupying the public easement and dealing with litigation as it arises. Another common example is the installation of conduit or wires within the public easement connecting two noncontiguous lots owned by the same person. The title attorney should be watchful for such unauthorized encroachments, especially where one business is being conducted upon two noncontiguous lots separated by an intervening lot.

55. Communications Act of 1934, 47 U.S.C. §§ 521 et seq.; *see also* Cable Communications Policy Act of 1984, Pub. L. No. 98–549, 98 Stat. 2780.

56. Some courts speak of this as apportionment of a utility company's right.

57. Cable Investments, Inc. v. Woolley, 867 F.2d 151 (3d Cir. 1989); Cable Assocs., Inc. v. Town & Country Mgmt. Corp., 709 F. Supp. 582 (E.D. Pa.1989).

58. Centel Cable Television Co. of Ohio, Inc. v. Cook, 567 N.E.2d 1010 (Ohio 1991); Mumaugh v. Diamond Lake Area Cable TV Co., 456 N.W.2d 425 (Mich. Ct. App. 1990).

## 4. Mailboxes and Paperboxes

One common encroachment that has recently been brought to light by litigious parties is mailboxes and paperboxes residing on the neighbor's property.[59] Often at the direction of the postal carrier, landowners are told to erect their mailbox across the road from their driveway entrance. While it has been argued successfully that the mailbox is a legitimate public use within the public easement,[60] the argument would not be as persuasive regarding the paperbox that is often set next to the mailbox.

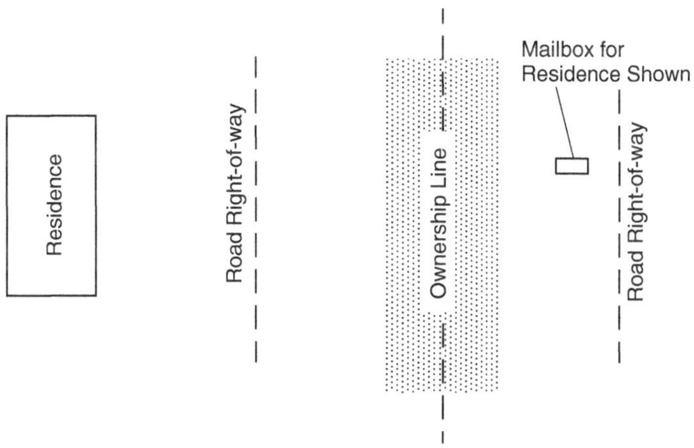


**Figure 13.17** The figure shows a common situation where the mailbox is across the public road from the owner's residence. Since the owner's boundary stops in the center of the road, the mailbox resides on the neighbor's land and may represent an actionable trespass.

Ordinarily, such minor encroachments are not shown out of ignorance or in order to avoid the appearance of frivolity. Nevertheless, the title attorney is cautioned to examine the situation carefully, especially where the mailbox or paperbox presents a nuisance or the neighbors are inclined to disagree over small matters.

## 5. Commercial Utilities in Public Roads

Almost all states by statute or court decision allow common and reasonably necessary utilities such as electricity, sewer, and water to reside within the public road easement. Ordinarily, such use passes without notice or comment by the surveyor. In fact, surveyors are more prone to comment where utilities are not found in a public easement. However, it must be recognized that some early grants for a public easement were limited in scope. A reasonable interpretation of the ancient grants would not allow utilities to share the road easement.

---

59. Miller v. Nichols, 526 A.2d 794 (Pa. Super. Ct. 1987).

60. In fact, in *Miller v. Nichols, id.*, the mailboxes were found to properly reside in the public easement. The case is controlling only within Pennsylvania and had a very convincing dissent that may be persuasive in other jurisdictions.

Example: " . . . to the town for road purposes only and no other purpose whatsoever without the grantor's permission . . ."

Obviously, such restrictive wording in the ancient grant of a public road easement may present a serious title problem where utilities subsequently share the easement with the public road. This possibility is often ignored with good cause. First, the maintenance of an action to remove the utilities will likely earn the ire of reliant neighbors and friends who would be cut off from service should the landowner succeed in litigation. Second, it must be conceded that such an obstinate position often results in the landowner "cutting off **[HIS]** their nose to spite **[HIS]** face." In other words, any successful claim will deny the utility company the very conduit the landowner relies upon to get utility service. Third, the continued trespass over a long time period has in almost all cases been maintained for sufficient time to give cause to a prescriptive right.

### *D. Not Shown*

A commentary on title problems that are disclosed by the surveyor would not be complete without explaining that certain situations affecting the title may be present and yet not shown. In other words, the absence of a notation on survey documentation does not necessarily mean the problem does not exist or that there should be no cause for concern or further investigation. Most surveyors lack the expertise, experience, and training to locate some potentially problematic areas. In other cases, surveyors do not believe it is within the scope of their professional duty to identify and locate certain items such as floodplains, designated agricultural lands, protected wildlife habitat, hazardous wastes, and the like. Therefore, the failure to show or mention these areas or topics is not meant to imply that they are not present. The following subjects typically escape review and detection by surveyors:

a. Inchoate Mechanic's Liens—The surveyor does not enter the dwelling to perform surveying services. Furthermore, even if there is evidence outside of the building that suggests inchoate mechanic's liens, most surveyors will not note it.
b. Wetlands—Most surveyors do not have the training, experience, or education to properly identify wetlands. As a result, wetlands (especially marginal wetlands) are not ordinarily shown on surveys—even though some statutes expressly require all wetlands be shown on subdivision plans.
c. Designated Agricultural Land[61]—Surveyors will not ordinarily identify or show nearby designated agricultural land.

61. Designated agricultural lands are lands that have actively sought special designation and are therefore immune from tortious claims or nuisances brought about by common and reasonable farming practices such as confining animals or spreading manure, weed killer, insecticide, or fertilizer on the fields and crops.

d. Floodplain—In most cases, the location of the Federal Emergency Management Agency (FEMA) one hundred-year floodplain limit is not located or shown.
e. Proper Subdivision—Surveyors assume that attorneys will review the lot for proper subdivision approval, though most attorneys are not familiar with subdivision requirements. As a result, violations often escape notification or detection.
f. Zoning—Most surveyors will only review setback distances. Therefore, nonconforming uses and other violations would not be discovered. In some cases, surveyors will show the distance from the improvement to the boundary and assume the attorney will compare the distance shown with the legal setback distance. A few surveyors will simply call the zoning or code enforcement officer and see if that official is personally aware of any violations.
g. Below-Surface Improvements—Unless there are depressions or other surface indications, below-surface improvements, such as the septic system leach field, underground utilities, drains, and so on, are not located by the surveyor and consequently cannot be evaluated as to their effect on the title or proposed use of the property.
h. Hazardous Waste—Surveyors do not attempt to look for or identify hazardous waste on the property. Many surveyors would miss even obvious signs of hazardous waste such as discarded drums, areas of dead vegetation, unusual discoloration of the soil, discarded asbestos insulation, and so on because they have no training or experience in the identification of hazardous waste.
i. Grave Sites—Unless the graves are located on or immediately adjacent to the boundary, the surveyor would not ordinarily discover grave markers or other indications of human burial. It would be rare for a surveyor to discover ancient Indian grave sites or marked graves in the interior of large parcels.
j. Special Zoning—Zoning such as shoreland zoning is not ordinarily identified on the plan prepared by surveyors. Only major bodies of water will be shown and only if they exist on the premises being surveyed—even though special zoning may arise from less-substantial physical features.
k. Wildlife Habitat Area—Although environmental resource agencies have or are in the process of preparing maps showing protected wildlife habitat areas, surveyors do not ordinarily show these designated areas on their map.
l. Violation of Covenants—Surveyors will likely note violation of covenants if they are physical in nature and based on locative aspects (e.g., building setbacks). Covenants involving the use, structure, size, height, number, etc. of improvements will not likely be discovered or identified by the surveyor.

## IV. Conclusion

In conclusion, the prudent attorney should always advise his client about the advantages of obtaining a current boundary retracement survey before purchasing or improving real estate. Without doubt, a quality boundary retracement survey will be costly—not obtaining a survey may be more so. This article was intended to provide some guidance in evaluating historical surveys and descriptions and in utilizing surveying services for evaluating title. No attempt was made to describe how to choose a surveyor or what to seek in contracting for surveying services. There have been other articles written on the topic.

The words of Judge Robert Gillespie may be most appropriate in reinforcing the need and value of having a boundary retracement survey:

> [L]and line cases usually generate a lot of heat and sometimes violence. These disputes involve the ownership of strips of land often only a few inches wide. For one reason or another, the narrower the strip of land in dispute, the more intense the feeling between the parties. Also, the less the property in dispute is worth, the greater the hostility. Before the case reaches the lawyers, angry words have usually passed between the parties. Sometimes a gun has been drawn, and before the case is over, a fight is not unusual. Now and then a party takes a "pot-shot" at his adversary; homicides are not unknown. Brother has slain brother; kin have fought kin; and neighbor has killed neighbor. It sometimes happens that the best and most reasonable citizens becomes involved in a land line suit, and nothing herein is intended to criticize or poke fun at such unfortunate persons. No man should be criticized for defending his rights.[62]

62. Justice Robert G. Gillepsie, *Some Animadversions on Land Line Cases*, 33 Miss. L.J. 151 (Mar. 1962).

# Coordinating Land Surveys in Multiple Jurisdictions

# 14

Landon Blake

## I. A Couple of Definitions

For the purposes of this article, a *land survey* is a bundle of professional land surveying services provided by a licensed land surveyor in accordance with a written scope of professional services for a specific project. Typically, the land survey produces one or more deliverables, or products, such as a map of property boundaries, a written description of land, or a topographic survey that is provided to an engineering professional or architect. A *jurisdiction* refers to a geographic area governed by its own set of land surveying and land development regulations. This could be a municipality, a county, or a state. The licensing of land surveyors, like that of lawyers, is done on a state-by-state basis. Therefore, when we refer to a jurisdiction in this article, we will typically be talking about a state.

## II. Why Does the Need Arise?

Why does the need arise to coordinate land surveys in multiple jurisdictions? There are many circumstances that may require it, but this article will briefly consider two of the reasons. The first and most common reason that organizations need to coordinate land surveys in multiple jurisdictions is in connection with an expansion of their operations. One example would be a specialty food grocery chain expanding with new stores in urban centers around the Western United States. The grocery chain may require topographic and boundary surveys of each new store site. The topographic surveys will be used to design the physical improvements for each store. The boundary surveys will be used to (1) identify title problems, (2) verify the setback requirements for buildings and other physical improvements from parcel boundaries, and (3) assist in securing loans in which the parcel will be collateral.

The second reason organizations may need to coordinate land surveys in multiple jurisdictions is to manage property that is (1) adjoining a boundary between jurisdictions or (2) on both sides of a boundary between jurisdictions. One example would be a large farming operation that owns several large parcels on both sides of a boundary between two counties. Any surveyor executing a boundary survey of the land owned by the farming operation would need to be familiar with and comply with the ordinances related to land surveying in both counties. (In California, any record of the boundary survey placed in the public record must be reviewed by the county surveyor in each of the counties.)

## III. The Lawyer's Role in Coordinating Land Surveys in Multiple Jurisdictions

A lawyer may be asked to perform several tasks as part of this coordination. These tasks may include the following:

1. Seeking statements of qualifications from licensed land surveyors.
2. Defining a requested scope of professional services as part of a response to a request for proposal (RFP) or contract negotiations.
3. Ranking survey proposals.
4. Awarding a contract for survey work.
5. Reviewing survey deliverables or products.
6. Ensuring survey products satisfy the requirements of other consultants, government agencies, and the contractor.
7. Determining how to handle problems or conflicts that are discovered during the execution of surveys.

The information and strategies in this article are designed to help lawyers handle these tasks.

## IV. Challenges to Coordinating Land Surveys in Multiple Jurisdictions

There are a number of challenges to coordinating land surveys in multiple jurisdictions. Some of these challenges relate to the different legal systems that govern land surveys and land development. As an example: One state may only allow civil engineers to create grading plans for land development projects, while another state may allow land surveyors to do this work. In another state, civil engineers may be allowed to conduct topographic surveys, while others may require that this work be performed by a land surveyor. As another example, some cities will require that a land surveyor set property corner monuments and file a public record of his survey if he adjusts a boundary between parcels. Other cities will not require this.

Other challenges to coordinating land surveys in multiple jurisdictions relate to the required interaction with more than one licensed land surveyor or land surveying company. Standards of practice can vary greatly between land surveyors. The quality and format of survey deliverables produced by each land surveyor can also be different. There can also be big differences in the hourly rates and fees charged by land surveyors. This is true even among land surveyors that operate in the same geographic region.

I should note that most licensed land surveyors understand the obligations, responsibilities, and pressures on the licensed professionals involved in any land development project. This understanding will hopefully make the working relationship between the lawyer and the licensed land surveyor on the project a positive one.

## V. The Options for Coordinating Land Surveys in Multiple Jurisdictions

There are three basic options that we will consider in this article. In the first option, the lawyer hires a multistate consulting firm to perform all of the surveys the organization will need. Although there are a few large, multistate land surveying companies, most often these consultants offer land surveying and related engineering services. These companies often charge higher hourly rates than smaller companies to cover their higher overhead costs.

These larger companies may also lack the local knowledge and expertise that small surveying companies possess which would benefit your particular project. For example, a small company with a limited area of practice may be familiar with local boundary conditions and local boundary research sources. They may also have good working relationships with staff at local public agencies, utilities, railroads, and timber companies.

In the second option, the organization directly contacts and contracts with local land surveying companies in each jurisdiction. This provides you with the benefits of having local companies on the project, but adds the headaches of dealing with multiple providers for your surveying services. Organizations that choose this option must be prepared to standardize the scope of professional services and deliverable items, to the extent possible, for all of the land surveyors providing services. There will be an additional cost for the organization to manage the multiple surveying consultants.

In the third option, the organization hires a single company providing land surveying services to supervise and coordinate the work done by other land surveyors working on the project. In most cases, the coordinating land surveying company will not contract directly with other surveyors, but will ask that the contracts be with the company. However, this coordinating land surveying company can take care of releasing an RFP, screening qualified consultants, defining a standard scope-of-work, standardizing project deliverables, and arranging for contracts to be negotiated and signed. They can also fill the role as an advocate for the organization that understands land surveying and the role that land surveying services will play in the successful completion of the project.

All three of the options described have their advantages and disadvantages. Of the three options, I favor the third. Careful selection of a coordinating land surveying organization will result in superior project performance by all land surveyors involved in the project. This superior project performance includes survey deliverables of superior quality, fewer schedule delays, and fewer arguments about the scope of professional services. Ultimately, all of these factors lead to a better product for a lower fee. The land surveyors working on the project will appreciate speaking to a client representative (your coordinating land surveyor) who speaks their language, understands their risks, and can acknowledge the challenges they will face in their work.

## VI. Negotiating the Scope of Professional Services

There is no other part of your role as an attorney when coordinating land surveys in multiple jurisdictions that is more important than negotiating a scope of professional services. A well-defined scope of professional services will ensure that you get everything you need to complete a successful project while also ensuring the land surveyor is compensated for all of his work and risk. A poorly defined scope of professional services can result in missing survey deliverables and lengthy and heated arguments about what was included for the price agreed upon for surveying services.

A scope of professional services is a written document that defines the services to be performed by a professional hired as a consultant to work on your client's project. In our case, the professional is a land surveyor.

The scope of professional services is a vital part of any agreement to hire a land surveyor, and you should not consider bringing a surveyor to your project team without one. For simple projects, the scope of professional services can be brief, while more complex projects should have a more detailed scope. Most scopes of professional services are organized into a set of individual tasks. Each task should come with a description, a set of understandings about the task, a set of exclusions for the task, and a specific set of deliverables for the task. (It may be appropriate to include a global list of understandings and exclusions at the end of a scope of professional services as an alternative.)

For example, you may have a topographic surveying task included in the scope of professional services for the surveyor. The description for this task would include the "what and where" being included in the topographic survey. (The portion of the 42-acre Sandy Creek Business Park lying south of the railroad tracks and west of the creek.) The understandings for this task could include the requirement that the client provide access to the property and a stipulation that the delivery of the topographic surveying data will depend on the weather's impact on field surveys. The task would also list exclusions to the topographic survey, such as identification of tree species and mapping of underground utilities.

A separate article could be devoted to discussing important topics about negotiating a scope of professional services with a land surveyor. This includes how to allocate risk, how a phased scope of professional service can reduce risk, common exclusions to a scope of professional land surveying, and others. In this article, however, the focus will be on a single concern related to negotiating the scope of professional services for surveys of multiple sites and/or surveys in multiple jurisdictions.

It is important to understand that the scope of professional services for land surveying must usually be tailored to each jobsite and jurisdiction. It is difficult to create cookie-cutter plans when the scope of professional services involves multiple jobsites. If the client attempts to apply the same scope of professional services to multiple jobsites, the responding land surveyors will have to raise their fees to mitigate the corresponding increase in risk. (It is possible to start with a template scope of work that is modified by jobsite or jurisdiction, as needed.)

A brief example will illustrate this issue. Return to the specialty grocery chain we mentioned earlier in this article. Ultimately, the grocery chain has the same basic goal for every jobsite: a new grocery store. They may even have a standard store design that will be used to build each new store. However, the requirements to build a store on each jobsite could be very different. One of the jobsites may need to deal with flood hazard issues, while another may need to perfect a land subdivision or lot line adjustment. A third may require a zoning variance. These types of issues will all require modifications or additions to the land surveyor's scope of professional services.

The different requirements of each jurisdiction may also require changes in the scope of professional services provided by a land surveyor. One city may require a site plan for your project, while the city down the road may not. Some states may require that your land surveyor set monuments to mark the boundaries of your new parcel, while others may not. These differences can result in significant changes to the scope of professional services performed by your land surveyor in each jurisdiction, with a corresponding difference in price.

What is the best way to handle the challenges to a uniform scope of professional services when coordinating multiple surveys in this manner? Create a template document that can be used as a baseline. Modify the scope of professional services as needed, or allow the land surveyors responding to your request for proposal to do so. When a land surveyor calls with questions about your project or the request for proposal, listen carefully. Changes to the scope of professional services may reduce the risk for the land surveyor and result in a lower fee for your organization.

We will conclude this section on negotiating a scope of professional services with a word about project deliverables. It is critical that the scope of professional services and the list of deliverables provided by the land surveyor meet the requirements of the project. This list of deliverables can often be difficult to determine if you do not have a strong land surveying background. This is one benefit of hiring a land surveyor to coordinate the surveying services for your projects, an option recommended above. You can utilize this land surveyor to determine an objective and a comprehensive list of deliverables that will be needed for each one of your jobsites and in each of the jobsite jurisdictions. As an alternative, select a single land surveyor based on qualifications, not on price or scope of professional services, and allow that land surveyor to guide you through the process of defining the deliverables. A reputable and high-quality firm will make sure you define the deliverable you need to have a successful project. You can find land surveying organizations that are willing to protect your interests when they negotiate their scope of professional services. However, these are not typically the organizations that offer the rock-bottom price. Remember that you get what you pay for.

## VII. Wrapping Up the Project

There are two tasks that must be completed as the land surveying portion of the project comes to completion. These tasks should be completed on any project in which land surveying is involved, but there are some special considerations when working on surveys in multiple jurisdictions.

The first task is a review of the deliverables for each jobsite. These deliverables could be items such as a topographic survey delivered to your civil engineer in digital format, a completed FEMA elevation certificate, or a filed map showing the location of the parcel boundaries on your side. Your review of each deliverable should ensure it meets (1) your client's needs, (2) the needs of other consultants working on the project, and (3) the requirements of public agencies that will review and approve the deliverables.

The second task is to ask your land surveyor questions about the work performed on your projects. These questions are important because the land surveyor works in an imperfect world. You need to understand the limitations of the land surveying work performed and what its implications are for the project.

When the project is complete, you want to have your land surveyor answer these questions:

1. How was the surveying control for the field surveys at each jobsite established? What horizontal and vertical datum were used? How does this compare to other datum used in the jobsite area and with datum used for previous work on this jobsite?
2. What equipment and methods were used to make field measurements as part of your survey for each jobsite? How was this data processed and adjusted? What were the limitations of this equipment and these methods?
3. What record sources were used during research for any boundary determination performed on each jobsite? Were there record sources that you were not able to search or records you were not able to locate? What key documents were used in your boundary determination and what is their quality? Were the documents for parcels adjacent to the jobsite examined? Did the information they contained agree with your boundary determination for the jobsite parcel(s)? Is there additional boundary research or surveying that could be performed to produce a higher-quality boundary determination on these jobsites? Are there likely alternative boundary determinations for these jobsites?
4. Did site conditions match existing construction plans, surveys, or as-built drawings?
5. Did physical occupation (fences, walls, and so forth) agree with your boundary determination?

Remember that the answers to these questions will differ depending on the jobsite and the jurisdiction. I recommend you ask for a written surveyor's report on each jobsite. This report should include answers to the questions listed above and should be signed and sealed by the land surveyor responsible for and in charge of the work described in the report.

## VIII. A Few Last Words on Price

I want to close this article with a few words on the price of land surveying services. The money spent preparing proposals and pursuing projects is a significant cost for most land surveying companies. If you properly package and offer your RFP for land surveying services on multiple sites, you will likely be able to realize a cost savings. (The land surveyor can discount his fee because he will be receiving more work for less proposal costs.) This cost savings will be negated if you make each land surveying company bid on each site individually. If you want to solicit multiple proposals to ensure you receive competitive fee estimates, award multiple jobsites to a single company, but ask for bids from more than one land surveying company. (If you have many sites, you could still award the services to multiple companies, but award the sites in small groups so you still get the discount that results from lower proposal costs.)

If you are going to ask multiple land surveying companies, make this clear in your RFP. It is even better to tell the companies that are considering a response to your RFP how many and which companies you are seeking bids from. This allows them to make an honest appraisal of the likelihood of a contract for the proposal effort required. The time of your land surveying peers is valuable, and they will appreciate your honesty.

If you do not work for a public agency and there isn't a requirement for multiple bids, I strongly encourage you to establish a relationship with a local land surveying firm. If they are a quality company and they know you are a repeat client, they will discount their fees to keep your business. They can do this for the same reason mentioned above: lower proposal costs. Most land surveying companies are willing to go the extra mile for repeat clients. Clients that simply seek the low bidder will find that they receive the poorest service.

If you properly package a well-written RFP, you can more than offset the cost of working in multiple jurisdictions by reducing the land surveyor's proposal costs.

When you are shopping for land surveying services, remember that you get what you pay for. You wouldn't select a doctor because he or she had the lowest fees, and the same applies to a land surveyor or any other professional. Your client will pay less in the long run if they hire a quality company for a reasonable fee.

# Reading, Interpreting, and Writing Land Descriptions

# 15

Donald A. Wilson, LLS, PLS, RPF

## I. Introduction

One of the most common problems shared by lawyers, surveyors, and others dealing with land records is that of land, or land-related, descriptions. There are more deeds and related documents than in almost any other category of concern, and therefore not only more rules and guidelines governing them, but also more potential for confusion, ultimately resulting in more disagreements and more litigation.

When the word "description" is mentioned, one immediately thinks of a deed, and often the two are equated. However, they are not one and the same. Deeds *contain* descriptions, but so do wills and other probate records, especially partitions, grants and patents, purchase and sales agreements, mortgages and foreclosure instruments, eminent domain proceedings, and town line perambulations and maps—the latter especially where dedicated easements are concerned. Generally speaking, the same rules apply to all, equally.

Since deeds and related title documents are a form of contract, basic contract requirements include the identification of the subject matter of the contract. In a deed conveying land, the description of the property to be conveyed must be such as to identify it or afford the means of identification, aided by extrinsic evidence.[1]

## II. Types of Descriptions

*Metes and bounds. Mete* means to measure, while *bounds* means the limits or boundaries of a tract of land.[2] While there have been a number of types of descriptions placed in this category, strictly speaking a *metes-and-bounds* description is one that has a defined point of beginning and travels around the parcel in a clockwise or counterclockwise direction, usually reciting courses, distances, monuments, and abutting parcels as they are encountered, returning to the point of beginning. Generally, an area of the parcel so traversed is recited. The theoretical parcel will contain all of the above elements; however in reality, there is a combination of the foregoing, with information lacking in many instances.

The *metes-and-bounds* description is commonly found in the eastern part of the United States and Canada, termed the "metes-and-bounds states," and is often used in present day. Smaller, irregular parcels lend themselves readily to this type of description, explaining its frequency in certain circumstances within the U.S. Public Land Survey System. Certain types of land parcels within that system have traditionally been described this way, as demonstrated in figures 1, 7, and 8.

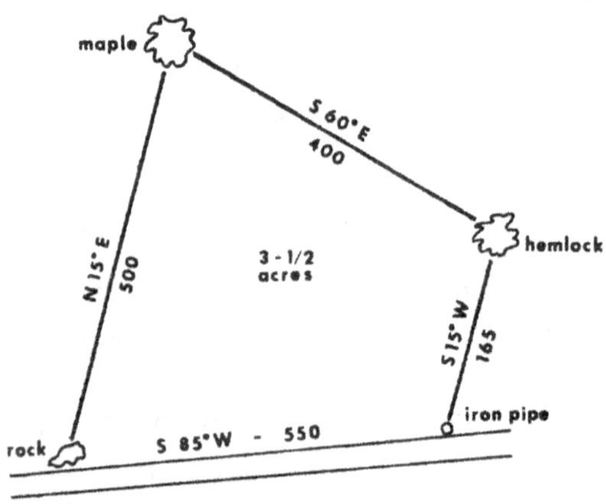


**Figure 15.1** Descriptions by metes and bounds
—From Donald A. Wilson, Deed Descriptions I Have Known but Could Have Done Without

*Adjoiners.* Also known as *reference to adjoining lands,* a description by adjoiners is one that calls mainly for abutting ownerships, although it may have an

1. City of North Mankato v. Carlstrom, 212 Minn. 32 (1942).

2. 11 C.J.S. *Boundaries* § 4. Metes and bounds means the boundary lines or limits of a tract. Metes may be understood as the exact length of each line and the exact quantity of land in square feet, rods, or acres. Metes result from bounds; and where the latter are definitely fixed, there can be no question about the former.

occasional corner monument call. As such, the abutting parcels are considered to be monuments.[3] While insuring that there are no gaps or overlaps, the *abutting description* inherently has two problems associated with it. The first, and perhaps the least troublesome although requiring considerable effort, is that the abutting ownerships called for are usually many, often a hundred or more, years out of date. This means that the current owners of abutting parcels must be researched back to that point in time where it is certain that they were owned by the persons called for. Only by doing that can one insure that the calls are satisfied.

The other problem results in a matter of interpretation. When an *abutting description* is interpreted literally, there can be several possible outcomes. Generally, however, when drawing a picture of the parcel, a strict literal interpretation should be taken. Refinement can be done later as more information is gathered.

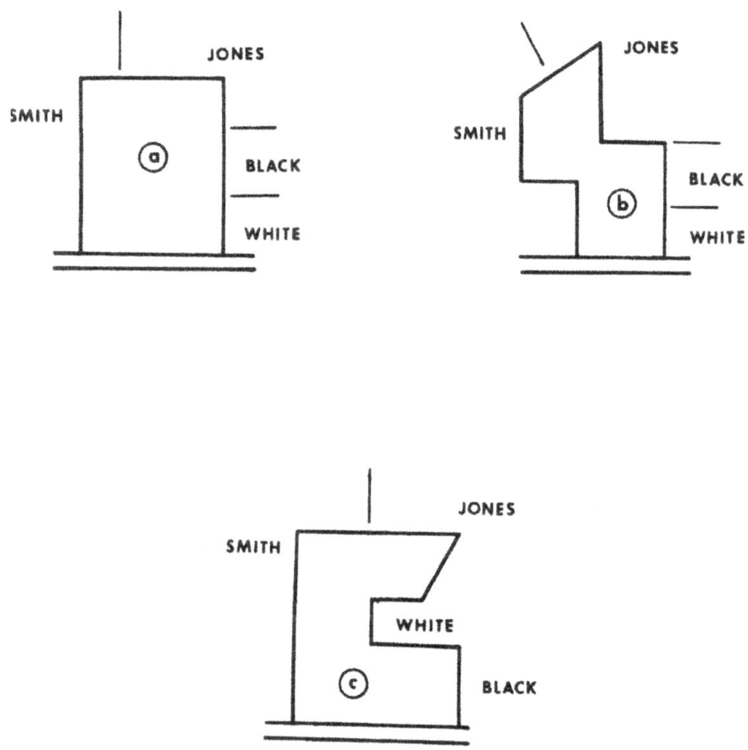


**Figure 15.2** Descriptions by abutting ownerships
From Donald A. Wilson, Deed Descriptions I Have Known but Could Have Done Without

*Lot and range.* A number of towns and townships were originally laid out in regular design, usually in rectangles and sometimes in squares or parallelograms, with an occasional triangle. Initial conveyancing, particularly original granting to

3. 11 C.J.S. *Boundaries* § 53. Where abutters are certain, they are monuments of the highest dignity. Adjoiners, in order to have controlling effect, must be established and well known, and must be called for in the conveyance; and where such calls are manifestly erroneous, they will be disregarded. When land is called for, it must be run to regardless of course and distance.

proprietors, consisted of original lots, or parts thereof. Subsequently, depending on the area and extent of conveyancing, these lots have been further subdivided into regular parts thereof, or into smaller parcels described by metes and bounds. In some areas, descriptions remain in terms of lots or parts thereof.

The complicating factor with this type of description is that the original lotting scheme must be known since each town was lotted independently, and each one is likely to be different from any of the abutting towns. Many are unique in design as there was no set pattern. Design, number, and arrangement of lots were governed by the shape of the town, the terrain, the number of original grantees or proprietors, and the instructions given to the surveyors or the "lot layers." It is therefore essential that one have a copy of the original lotting plan whenever possible. When there is no plan available, one must be reconstructed from available evidence of surrounding descriptions and physical evidence. Occasionally it is necessary to reconstruct an entire town or a large portion thereof in order to determine the scheme.

Towns were often laid out in *divisions,* which was a planned scheme to take advantage of and grant accordingly the quality of the land. There are frequently several divisions in any given town, so any lot number may appear several times, once in each division. There was also common land left in many towns for later granting, compensating for lot deficiencies or for the use of the public. Some of this was later granted, or lotted, indiscriminately. The term "pitch" is used for this procedure in some locations.[4]

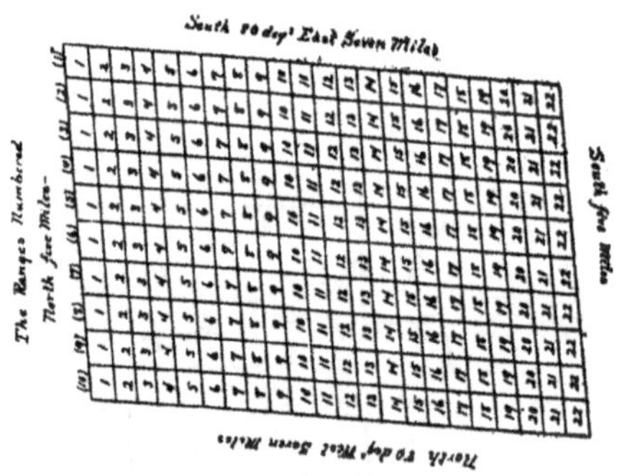

**Figure 15.3** Example of a township laid out in lots and ranges

*Plan reference.* Some descriptions merely refer to a plan (commonly termed "plat" where a subdivision of land is concerned), which by rule of law makes the

4. "Pitch" is sometimes used for undivided land in town lottings which are not part of the regular divisions. There were a number of ways by which occupancy and title evolved, the most common of which was to find an existing bound at the edge of the undivided land and survey a parcel out the pitch, for recording. The person surveying, claiming, or "improving," and recording the pitch and paying the taxes on it became its owner, similar to homesteading.

plan part of the description.[5] Therefore, the plan must also be consulted in order to determine the location of the parcel as well as the definition of its lines and corners. This is most important since there is usually an abundance of information appearing on the plan that is not included within the deed.

The problem arises with this type of description when the plan is not on public record, or worse, when it cannot be found.[6] Sometimes a lengthy search is necessary to locate such a plan, and because it may give additional information or information contrary to the deed description (e.g., if the description was written incorrectly from the plan), its importance cannot be overemphasized.

This is a very common type of description in modern conveyancing, where there have been numerous subdivisions of land under the various laws and regulations and then descriptions written from those plans. Because extreme care has not been taken in all cases, there is often incomplete information in the deed or conflicting statements between the deed description and that appearing on the plan.

*Block and lot.* Many of the early municipal surveys and lottings, and some modern subdivisions—particularly larger ones or those with several phases of development—have been designed with lots within blocks, or sections. Great care must be taken when dealing with a lot number since that particular number may apply to several lots, depending on which block or section is concerned.

Some include other forms of description under this category: arpent lotting, lot and range, townsites, and occasionally other variations. The common denominator is that they all require a map, plan, or plat to show the lot and its particulars.[7] Problems arise when such map cannot be found or is lost. In such situations, sometimes a large area, or the entire subdivision, must be reconstructed to determine what is being described.

*Coordinates.* Surveying computations and much design are accomplished through the use of *coordinates* or with a *coordinate system.* Mathematically, all points are designated with a north and an east coordinate value based on a defined point of origin. The origin may be "local" or arbitrarily selected for the particular project, or it may be a part of a widespread system of coordinates.

---

5. Gibson v. Johnson, 244 So.2d 713 (La. Ct. App. 1971). When a description refers to a plat, the plat itself, with all its notes, lines, descriptions, and landmarks, becomes a part of and controls the grant. Ford v. Ward, 130 So. 2d 380 (Ala. 1961). Lines and figures are part of the description as if the courses and distances were set out in the deed. Dolde v. Vodicka, 49 Mo. 98 (1871). Where no other description is given of lands sold than the number of the lot in a survey of a tract of land, or the plan of a town, or an addition to the same, the authentic map of such survey is as much a part of the deed as though set out in it.

6. Perkins v. Jacobs, 129 A. 4 (Me. 1925). A plan incorporated in a deed by reference need not be recorded, but it is sufficient to prove it and its contents. N.H. Land Co v. Tilton, 19 F. 73 (C.C.D.N.H. 1884). A valid deed does not become void because, by reason of the loss of a plat referred to therein, it has become difficult to define the boundaries.

7. Ford v. Ward, 130 So. 2d 380 (Ala. 1960). Where no other description is given of lands sold then by number of the lot in a survey of a tract of land, or the plan of a town, or an addition to the same, the authentic map of such survey is as much a part of the deed as though set out in it.

Some larger cities have their own coordinate system while each state has its own State Plane Coordinate System defined by statute. Conversions may be made between systems with the appropriate mathematical routines, and the values of such systems are (1) the ease of manipulation of numbers plus (2) the assignment of a unique position to each point. So long as the mathematics and the basis of any system are known, point positions are permanent and may be reconstructed at any time.

This system is technically superb but does not lend itself well to description in deeds and related documents except as supplemental data to traditional information. Providing coordinate values would be helpful to a surveyor but would mystify all but the most sophisticated landowner or deed researcher. The other problem associated with coordinates is the fact that a description or a plat with just coordinated points has no check on itself, whereas a description with bearings and distances can be computed for closure. A description with coordinated points will always close perfectly, and because of the high-tech aspects of the description, the reader is lulled into a false feeling of confidence.[8] In addition, if only coordinates are given, the average landowner has been robbed of his sense of security of having monumented corners or some set of directions to follow in order to find his lines and corners.

In addition, courts have traditionally taken a dim view of values derived from basic measurements, knowing they contain mathematical error. Directions and distances have long been treated with less dignity than all other calls, except for area or quantity. The Texas Supreme Court elaborated on the problem: "Course and distance are the most unreliable calls; distance is less reliable than course, because of the mistakes of the officers, over which the locator has no control. . . ."[9]

*Strip descriptions.* Used primarily for defining rights-of-way, this type of description generally contains a series of directions and distances defining a centerline, then a width of the strip or distances given for the width on each side of the centerline. Some minor easements, known as blanket easements, are less definitive in that they describe the right, but not its location or dimensions. A common example is where an easement is granted for a telephone line or a power line, and its exact location cannot be determined until after the line is installed: the easement describes the use, e.g., a telephone line, and the extent of the easement is the area where the line is actually built.

---

8. One example is where a transposition of numbers on an x-coordinate took place from 512,129.72 to 512,127.92 and the corner monument destroyed. A later restoration of the corner results in the monument being placed 1.80 feet too far to the west, but with a similar transposition on a distance from 129.72 feet to 127.92 feet, the mistake would be discovered and the appropriate correction made. James A. Thigpenn III, On Writing Descriptions (unpublished paper).

9. Stafford v. King, 30 Tex. 257, 257 (1867).

. . . conveying the right to install a line from the highway to the building(s) located on the premises at a convenient width and location. . . .

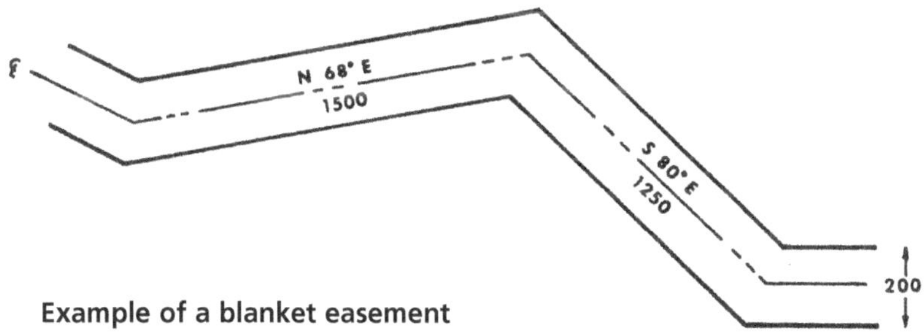


**Figure 15.4** Example of a strip description

*Arpent parcels.* Known in some locations as *river lots,* this type of arrangement is primarily of French origin and is frequently found along some bodies of water. They are prevalent in Louisiana and appear throughout the upper Midwest and in the Northeast, particularly in northern New England and eastern Canada.

Because roads were rare in the early development of the country, travelers and settlers relied on water, especially rivers, in traveling from place to place. The underlying concept of this type of lot arrangement seeks to provide frontage on the body of water and extend back, essentially at right angles, the necessary distance to provide the required acreage. This scheme resulted in a series of strips as shown in the diagram below. Descriptions were generally by lot number and seldom contain metes and bounds. Because of this, the reader must obtain a copy of the layout or, at the very least, have some knowledge of the particular scheme at that location.

*Area on one side of a described line.* Occasionally, a tract of land is described by no more than the area on one side of a described line dividing the parent parcel. Full knowledge of the tract being divided is necessary in order to properly apply the line so described. Usually the line in question will be defined by metes and bounds, though not always, but little if anything else describes the parent tract. In its simplest form, such a description may merely discuss a line that will sever a certain acreage, or a line that will divide the parent tract in equal parts.

Since this category includes the so-called "ly" description, problems can arise with irregular parcels since more than one interpretation may ensue. Great care must be exercised in interpreting this type of description, and it should be used only when there is no chance for more than one interpretation.

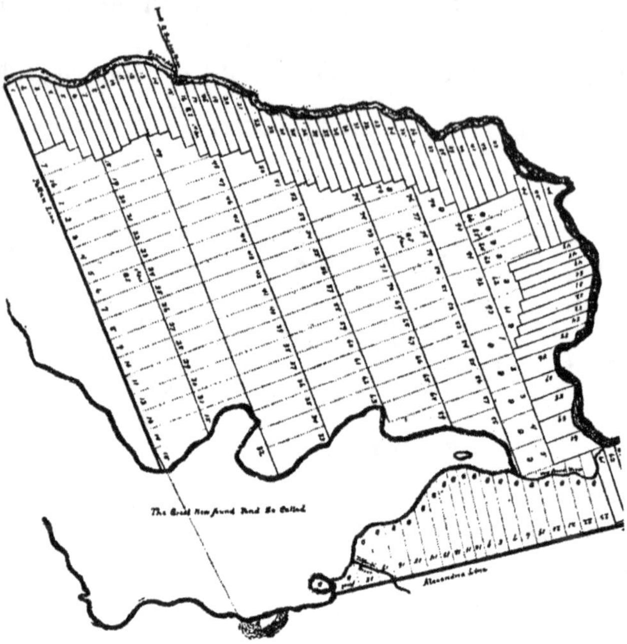

**Figure 15.5** Arpent parcels laid out along the Connecticut River in Hanover, New Hampshire

> . . . the West half of the Ridge Pasture so-called, the same being set off to the heirs of Loring Dunbar at the June term of the Probate Court. . . .
>
> Probate proceeding: . . . said pasture being divided as follows beginning at a stake and stones placed near the southern or front wall, ten rods six feet west from the southeastern corner running from the said stake and stones in a northerly direction to a pine tree marked on the south side from thence to a stake and stone place near the wall on the northern end ten rods six feet from the northeasterly corner of said pasture.

Example of a description of one line for the conveyance of parcels on either side of it

*Description by reference: (1) by name of occupant or owner.* Land occupied or owned by another party is sometimes difficult to identify with accuracy. Descriptions of ownerships can usually be found so long as they are on record or at least available, although they are not always easy to locate, especially if the title derives from an inheritance. Lands occupied usually present difficulty because they are seldom a part of the record in that ownership may be in another person, and relating the two, such as through rent or lease, may not be possible.

*Description by reference: (2) by name or number of tract.* A reference to a tract by name or by a tract number depicted on a map or plat may be a sufficient description to define the boundaries.[10] However, if the plat is unavailable, or deficient, problems may arise. The other frequent difficulty is that at some point, the parties may have known a particular tract of land by a local or colloquial name, which at that time carried some meaning but over time has been forgotten. A frequent encounter is the designation of a parcel with some previous owner's name, usually a long-term owner or prominent family from some time in the past. Researching back to that ownership is sometimes necessary to prove the correct parcel of land.

*Description by reference: (3) by description in other grant or conveyance.* It is not uncommon to find a description that merely refers back to a previous conveyance. By rule, any reference is as much a part of the description as if copied into it item for item, so any references must be examined.[11] Occasionally, referring to a previous document may merely refer the reader back to another document, and so on through several transfers over many years, to the original source. Such a description, however, may be perfectly adequate.

*Combination.* Many descriptions are combinations of the foregoing categories. Problems are inherent in many of these since there are frequent conflicts between description systems as well as conflicts between description elements. Reference may be made to "the relative importance of conflicting elements" for the resolution of many of these conflicts.

*Ranchos.* A number of Spanish land grants were made in some southeastern and some western states prior to statehood. Many of these were quite large and were honored as senior when rectangular surveys later came about. They are generally of the "metes-and-bounds" type, although often rather crude.

## III. Descriptions in the Public Land Survey System

While any of the foregoing types of descriptions may occur within the U.S. Public Land Survey System (PLSS), the following are confined to being within this system of rectangular surveys. One of the unique features of this broad category is that most surveying and description was done according to set rules and instructions, and therefore interpretation and retracement of descriptions fall within those rules and their supplements or amendments.

10. 11 C.J.S. *Boundaries* § 20c. A reference to a tract by name or by lot number as laid down on a map or plat may operate as a sufficient description to define the boundaries of land; and, where the number of the lot on a plan referred to is the only description of the land conveyed, the courses, distances, and other particulars are to be taken as if set forth fully in the deed itself.

11. 11 C.J.S. *Boundaries* § 20e. Where land is described in a grant or conveyance by reference to another grant or conveyance, the description contained in the latter is regarded as adopted by, and incorporated into, the former, and the land therein described will pass.

Land surveys and descriptions were generally, though not always, done according to a specific set of instructions. Modern treatises consist of the Manuals of 1947 and 1973 and recently the manual of 2009 along with, for many areas, Special Instructions.

*Aliquot parts.* The term "aliquot" means, strictly, contained in something else an exact number of times.[12] This type of description lends itself to descriptions including 1/2, 1/4, and the like, and many descriptions are merely that.

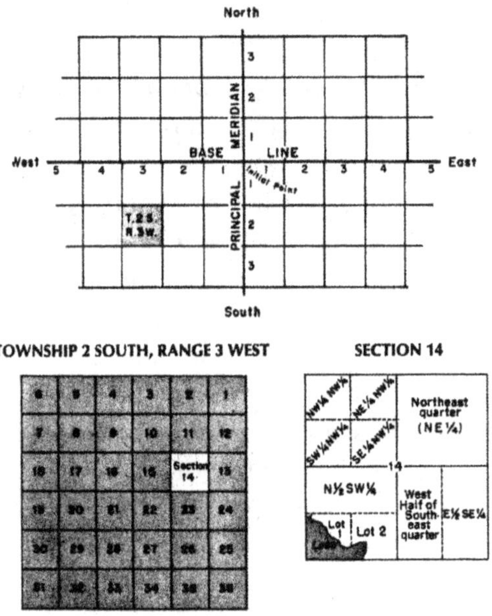


**Figure 15.6** Illustration of land division in the Public Land Survey (Rectangular) System

—From U.S. General Land Office, Restoration of Lost and Obliterated Corners (1883)

*Mineral surveys.* Mineral surveys exist where deposits occur in ore-bearing rock veins, known as mineral lodes, where the prospector who has made a legal discovery is permitted to stake out a claim of specified dimensions, to develop the same to his exclusive use, and, if desired in the protection of his right, to apply for a mineral survey and obtain a patent.

12. Black's Law Dictionary 81 (8th ed. 2004).

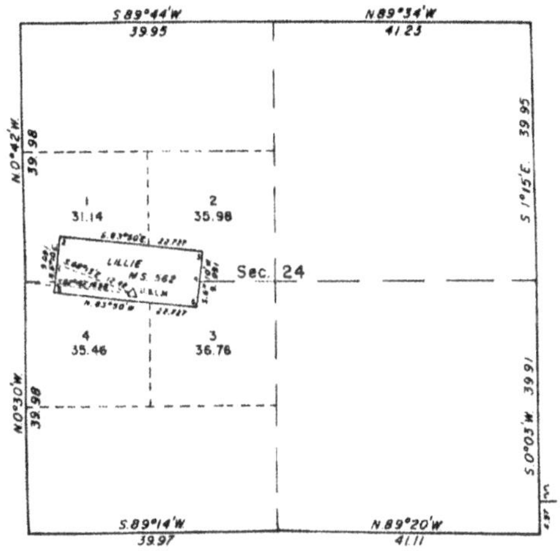


**Figure 15.7** Example of a mineral survey in Section 24
—From U.S. BUREAU OF LAND MANAGEMENT, MANUAL OF SURVEYING INSTRUCTIONS (1973)

*Homestead entry surveys.* Based on the Homestead Act of 1862, entries were made by individuals to settle land as homesteads. Where they exist, separate surveys may be found for them.

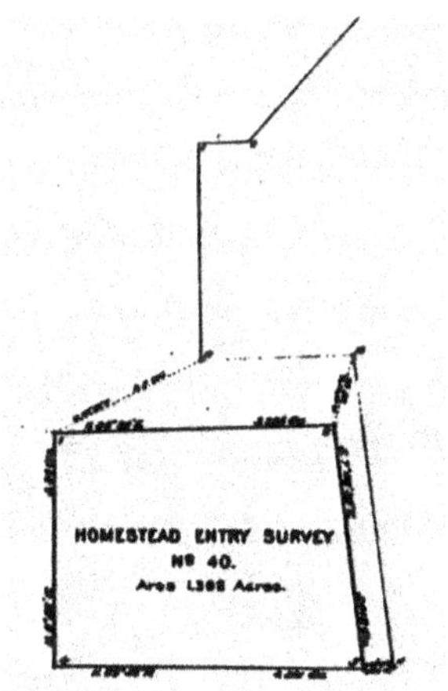


**Figure 15.8** Example of a Homestead Entry Survey
—From Notes, U.S. BUREAU OF LAND MANAGEMENT, ADVANCED CADASTRAL SURVEY TRAINING. Marana, Arizona (1983)

*Donation land claims.* There were Donation Acts in several of the states beginning in 1842. Where these acts applied, one generally will find surveyed descriptions of donated lands.

*Townsites.* A townsite survey, in public-land surveying practice, is a survey made within one or more regular units of the township subdivision by which the land is divided into blocks, streets, and alleys as a basis for the disposal of title in village or town lots. These fall into two general classes: those with few or no prior improvements and those where villages or towns already exist.

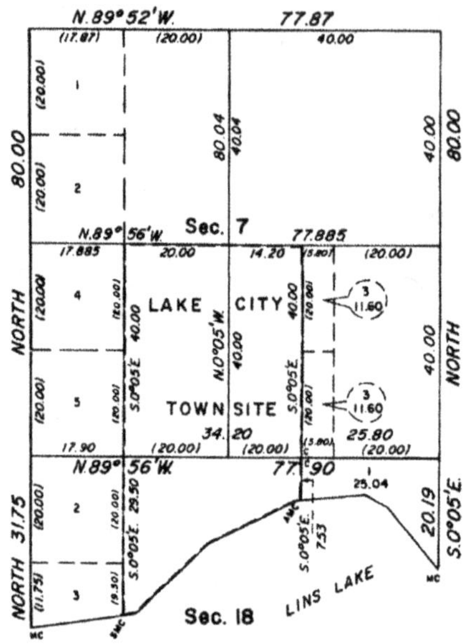


**Figure 15.9** Example of a townsite in Section 7
—U.S. Bureau of Land Management, Manual of Surveying Instructions (1973)

*Millsites.* Millsites embracing land occupied for milling purposes or used incidental to mining operations may be located, surveyed, and patented in a manner similar to lode claims.

## IV. Style of Description

With a "metes-and-bounds type" description, one typically encounters two styles of description. The *running description* basically "runs" around the property in either a clockwise or counterclockwise direction, reciting information such as directions, distances, monuments, and abutters as they are encountered. In a closed figure, the ending point will be the same as the beginning point, and the figure will "close" verbally. A *bounding description* is one in which the scrivener, or description writer, and ultimately you, the reader, is on the inside of the parcel looking out and describing its bounds.

As an example, the following parcel of land may be described both ways.

The running description would be something like:

> beginning at the southwest corner of the parcel being described, THENCE running in a northerly direction by land of Smith, 100 feet; THENCE easterly by land of White 150 feet; THENCE southerly by

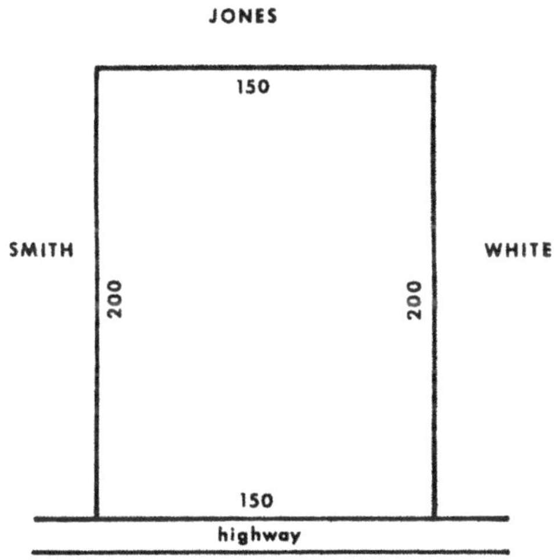


**Figure 15.10** Sample tract for description illustration
—FROM DONALD A. WILSON, DEED DESCRIPTIONS I HAVE KNOWN BUT COULD HAVE DONE WITHOUT

> land of Jones, 100 feet; THENCE westerly by the highway, 150 feet to the point of beginning.

A bounding description would read as follows:

> BOUNDED westerly by land of Smith, 100 feet; northerly by White, 150 feet; easterly by Jones, 100 feet and southerly by the highway, 150 feet.

If a running description is misinterpreted as a bounding description (or vice versa), the parcel will be drawn at 90° to what it should be. This may seem obvious, but caution is merited. Some descriptions are impossible to identify as either running or bounding descriptions from merely reading them alone; in such cases, to identify the type of description with accuracy, one must draw out the description and compare it with known features, such as the directions of roads or abutting parcels. Occasionally descriptions will be a combination of both types.

The best advice in reading descriptions is to assume nothing and take no chances. Always draw the description and compare it—even if only mentally—to known surroundings, features, and orientation. Following this advice should minimize mistakes.

## V. Sufficiency of Description

A deed will not be voided because some particulars of the description are false or inconsistent, so long as it is sufficient to identify the premises. However, if the description is so imperfect that it is impossible to know what land was intended to be conveyed, a deed or other instrument containing the description may be void. The basic rule is this: *if the description is certain enough to enable a person to locate the land, it is sufficient.*[13]

Courts have consistently ruled on the sufficiency of description when the lands can be located by a surveyor[14] or by extrinsic evidence.[15] In fact, the case of *Rivers v. Lozeau*[16] emphasized that "the definition of a legally sufficient real property description is one that can be located on the ground by a surveyor."[17]

## VI. Rules for Reading Descriptions

One should read a description a *minimum* of three times. The first reading will provide an idea of the parcel, and the reader can decide if it is the correct parcel of interest plus have an idea of its appearance and its orientation. The second reading should be item by item, drawing a sketch as the narrative proceeds around the parcel. The third reading is a check and verification of what was drawn, making certain that the drawing contains *everything in the description,* and *only what was in the description.* The resulting sketch should be as accurate as possible in accordance with the writing. Even if the reader does not agree with the description, and even if there are errors, he should put down exactly what the description states. If there are two interpretations, draw two sketches, one for each interpretation.

## VII. Rules for Interpreting Descriptions

*Rules of construction.* Where descriptions set forth in a deed are not ambiguous, they must be followed. Only when the meaning of a deed is not clear or is ambiguous or uncertain will a court of law or equity resort to established rules

13. Donald A. Wilson, Interpreting Land Records (2006).

14. Brunotte v. DeWitt, 196 N.E. 489 (Ill. 1935); *see also* Smiley v. Fries, 104 Ill. 416 (1882). Any description in a deed by which a parcel of property may be identified with reasonable certainty by a competent surveyor is sufficient. McCullough v. Olds, 41 P. 420 (Cal. 1895). If a surveyor, by applying the rules of surveying, can locate the land, the description in the deed is sufficient.

15. Persinger v. Jubb, 17 N.W. 851 (Mich. 1883). A description in a deed may be sustained if, with the aid of the usual deed inquiries, there would be no difficulty in locating the land. A deed will only be held void for uncertainty where, after resort to oral proof, it still remains a matter of conjecture what was intended by the instrument. Smith v. Crawford, 81 Ill. 296 (1876). The Mississippi case of *Peacher v. Strauss,* 47 Miss. 353 (1872), stated that "practically, no deed or conveyance of land was ever made, however minute and specific the description, that did not require extrinsic evidence to ascertain its location; and this is so, whether the description be by metes and bounds, reference to other deeds, to adjoining owners, water courses, or other description of whatever character." The court went on to say that "under every conceivable state of facts, and in every imaginable circumstance . . . contracts, wills and deeds are made effective by the identification of extrinsic evidence, of the person or subject intended, yet in no wise violating the rule, that such evidence 'cannot be admitted to contradict, add to, subtract from, or vary the terms of a written instrument.'" 47 Miss. at 358.

16. 539 So. 2d 1147 (Fla. Dist. Ct. App. 1989).

17. 539 So. 2d at 1150.

of construction to aid in the ascertainment of the grantor's intention by artificial means where such intention cannot otherwise be ascertained. Unlike a settled rule of property which has become a rule of law, rules of construction are subordinate and always yield to the intention of the parties, particularly the intention of the grantor, where such intention can be ascertained. Since all rules of construction are in essence only methods of reasoning, which experience has taught are best calculated to lead to the intention of the parties, generally no rule will be adopted that tends to defeat that intention.[18]

There is an abstract distinction between "construction" and "interpretation" in that "construction" is the drawing of conclusions from elements known from, given in, and indicated by the language used, while "interpretation" is the art of finding the true sense of the language itself, or of words or symbols. In other words, "interpretation" is used with respect to language or symbols themselves, while "construction" is used to determine not the sense of the words or symbols, but the legal meaning of the entire contract. However, in many contract cases, it would be difficult as well as unrewarding to determine where "interpretation" ends and "construction" takes over. It is doubtful if the abstract distinction between the terms is of any vital significance, and certainly, in common and legal usage, "construction" and "interpretation" are used interchangeably.[19] The *intent* of the parties is the controlling consideration.[20]

- In ascertaining the intention of the parties, separate deeds or instruments executed at the same time and in relation to the same subject matter, between the same parties, may be taken together and construed as one instrument.
- A deed must be construed as a whole and a meaning given to every part thereof.
- Words in a deed are presumed to have a purpose.
- Documents are to be viewed in light of the surrounding circumstances.

---

18. 23 AM. JUR. 2D *Deeds* § 221 (1983).

19. 17 AM. JUR. 2D *Contracts* § 240; 30 AM. JUR. 2D *Evidence* § 1065; WILSON, INTERPRETING LAND RECORDS, ch. 4 (2006).

20. The main object in construing a deed is to ascertain the intention of the parties from the language used and to effectuate such intention where it is not inconsistent with any rule of law. Iselin v. C.W. Hunter Co., 173 F.2d 388 (5th Cir. 1949). "Intention" as applied to the construction of a deed is a term of art and signifies a meaning of the writing. United States v. 15,883.5 Acres of Land in Spartanburg County, S.C., 54 F. Supp. 849 (W.D.S.C. 1944). The intent of the grantor as spelled out in the deed itself must be interpreted, not the grantor's intent in general, or even what he may have intended. Wilson v. DeGenaro, 415 A.2d 1334 (Conn. Super. Ct. 1979). In ascertaining the intent of the grantor in a deed, the court will not attempt to ascertain and declare what the grantor meant to say, but only the meaning of what the grantor did say. Holloway's Unknown Heirs v. Whatley, 104 S.W.2d 646 (Tex. 1937). Generally, secret intentions of the grantor not communicated to the grantee are not binding on the grantee. Cawthon v. Cochell, 121 S.W.2d 414 (Tex. 1938). A deed conveys the land actually described, regardless of the mistake of the parties. Land cannot be transferred by the intent of the parties alone, especially when the specific words used state less than what was intended. Tilbury v. Osmundson, 352 P.2d 102 (Colo. 1960). *See* WILSON, INTERPRETING LAND RECORDS ch. 4.

- Descriptions are to be construed according to their plain terms, and words are to be given ordinary meaning.
- A document should be construed in light of the law in existence at the time it was executed.
- A deed should be construed with reference to the actual state of the land at the time of its execution.
- Technical words should be construed according to their technical meaning.
- Words that have definite legal significance must be given their legal effect.
- A specific, or particular, description controls a general one.
- References are part of the description with as much effect as if copied therein.
- Descriptions are to be construed against the grantor and in favor of the grantee[21]
- A false or erroneous description or statement may be disregarded.
- A *meaning and intending clause* will not limit or enlarge the grant.
- *Punctuation* is ordinarily given slight consideration in construing a deed. Words will control punctuation marks rather than the other way around, however, punctuation will be resorted to in order to settle the meaning of an instrument after all other means fail.
- Courts generally take judicial notice of commonly used initials and abbreviations in descriptions.

*Relative importance of conflicting elements.* In determining the boundaries of a tract of land, it is not permissible to disregard any of the calls if they can be applied and harmonized in any reasonable manner.[22] In doing so, conflicts may arise within a description such that it is not possible to harmonize all of the calls. However, the courts have agreed upon a classification and gradation of calls in a grant, survey, or entry of land by which their relative importance and weight are to be determined. These rules are not artificial or based on theory but are the results of human experience. They are not binding requirements, or rules of evidence, or merely helpful guidelines in determining which of two conflicting calls should control, but *rules of construction,* that are adaptable to the circumstances of each case. A call that would defeat the intention of the parties will be rejected regardless of the comparative dignity of the conflicting calls, and, where calls of a higher order are made by mistake, the calls of the lower order may control as indicating the intention of the grant.[23]

---

21. This rule should be treated as a last resort, when all other rules fail to resolve the ambiguity. Smart v. Huckins, 82 N.H. 342 (1926).
22. 12 Am. Jur. 2d *Boundaries* § 66.
23. 11 C.J.S. *Boundaries* § 47.

The rules for the order of conflicting description elements are founded upon the principle that those elements are to control in which error is least likely to occur.

Relative Importance of Conflicting Elements

A. Natural Monuments
B. Artificial Monuments
   Maps, Plats, and Field Notes
   Adjoiners
   Metes and Bounds
C. Courses
   Angles
D. Distances
E. Area

## VIII. Corners and Marked Lines

Above all, it is the original survey that governs, hence giving rise to the familiar legal standard of "following the footsteps of the original surveyor." Recently, at least one state court has expanded on this requirement, stating that it is following the words of the original deed writer that is paramount[24] when there is no survey at the time of the creation of a land parcel. Robert J. Griffin described it a different way, stating as his first principle "The location of a boundary line is determined as of the time of its creation," [and that] "A boundary line once established should remain fixed in its original position through any series of mesne conveyances."[25] A grantee who purchases the entire extent of particular lands owned by the grantor determines the boundaries of his purchase as of the time that the particular parcel was carved out of some larger tract. He takes to the bounds of the estate of his grantor, who in turn took to the limits of his grantor's estate, etc., to the time of the creation of the boundary.

At times, tracing the chain of title of a parcel back to its creation will lend insight into its interpretation. Sometimes, enough additional wording or description is found, giving rise to correct interpretation and aid in avoiding misinterpretations.

## IX. Words and Phrases

Several of the foregoing principles that apply to the interpretation of land records depend on particular words and phrases. Many terms have been defined, and

24. Larsen v. Richardson, 2011 MT 195, 260 P.3d. 103.

25. *Retracement and Apportionment as Surveying Methods for Re-establishing Property Corners*, 43 MARQUETTE L. REV. 484, 487–488 (1960).

a number of them have been ruled upon by various courts. Those in the latter category may be found in a set of legal encyclopedias entitled WORDS AND PHRASES,[26] which contains most of the words and phrases a reader is likely to encounter. For others, either WEBSTER'S DICTIONARY or one of the law dictionaries such as BALLENTINE'S or BLACK'S will ordinarily suffice. For unique words or words not appearing in any of the above, special resources, which may consist of almost any imaginable reference source, must be consulted.

## X. Writing Descriptions

Once a person is familiar with how a description is to be read and interpreted, he or she will have insight as to how one should be compiled. Familiarity with basic writing skills is a prerequisite, and one standard text[27] has stated that the desirable qualities of a scrivener[28] are having knowledge of the following:

- Trigonometry, geometry, and all elementary mathematics
- The legal meaning of all words and phrases used in land descriptions
- The location of known controlling monuments in the area of the description
- The science of measurements and calculations to prove the correctness of measurements
- Senior rights of adjoiners
- The order of importance of conflicting elements

By understanding and following the appropriate rules and guidelines, the purposes for which descriptions are written will be better served and confusion will be minimized.

### *References*

Brinker, Russell C., and Roy Minnick, eds. 1987. *The Surveying Handbook.* New York: Van Nostrand Reinhold Company Inc.

Bureau of Land Management. 1973. *Manual of Instructions for the Survey of the Public Lands of the United States.* U.S. Department of the Interior. Bureau of Land Management Technical Bulletin 6. Washington: U.S.G.P.O.

Land & Boundary Consultants, Inc. Brochure entitled *Land Measurements.*

Thigpenn, James A. III. *On Writing Descriptions.* Unpublished paper.

Wilson, Donald A. 1981. "Descriptions by Abutters." *The Benchmark* (NHLSA) 6(4): 131–33.

Wilson, Donald A. 1980. "The Interpretation of Deed Descriptions." *The Benchmark* (NHLSA) 5(2): 9–13.

26. WORDS AND PHRASES, 100 vols. (West Publishing Co.).

27. BROWN, ROBILLARD, & WILSON, EVIDENCE AND PROCEDURES FOR BOUNDARY LOCATION (1994).

28. A writer; scribe; conveyancer. One whose occupation is to draw contracts, write deeds and mortgages, and prepare other species of written instruments. BLACK'S LAW DICTIONARY. 1375 (Eighth ed. 2004).

Wilson, Donald A. 1988. "Words in a Deed." *The Benchmark* (NHLSA) 12(2): 2–3.
Wilson, Donald A. 1982. *Deed Descriptions I Have Known but Could Have Done Without.* Rancho Cordova: Landmark Enterprises.
Wilson, Donald A. 2006. *Interpreting Land Records.* Hoboken: John Wiley & Sons, Inc.
Wilson, Donald A. 2008. *Forensic Procedures for Boundary and Title Investigation.* Hoboken: John Wiley & Sons, Inc.

## For Further Reference

Backman, James H., and David A. Thomas. *A Practical Guide to Disputes Between Adjoining Landowners—Easements.* New York: Matthew Bender & Company, Inc.
*Black's Law Dictionary*
Brown, Curtis M., Walter G. Robillard, and Donald A. Wilson. 2009. *Boundary Control & Legal Principles,* 6th ed. New York: John Wiley & Sons, Inc.
Brown, Curtis M., Walter G. Robillard, and Donald A. Wilson. 2009. *Evidence & Procedures for Boundary Location,* 6th ed. New York: John Wiley & Sons, Inc.
Cazier, Lola. Surveys and Surveyors of the Public Domain, 1785–1975. 1975. U.S. Department of the Interior. Washington: U.S.G.P.O.
Cuomo, Paul, and Roy Minnick. 1993. *Advanced Land Descriptions.* Rancho Cordova: Landmark Enterprises.
*Flick on Titles.*
Patton, Ruffod G., and Carroll G. Patton. *Patton on Land Titles.* 1957. St. Paul: West Publishing Company.
Wattles, Gurdon. *Writing Legal Descriptions.* 1976. Orange, CA: Wattles Publications.
Wattles, William. *Land Survey Descriptions.* 1974. Orange, CA: Wattles Publications.
*Words & Phrases.* 100 volumes. West Publishing Company.
White, C. Albert. *A History of the Rectangular Survey System.* U.S. Department of the Interior. Bureau of Land Management. Washington: U.S.G.P.O.

# A Guide to Gaps and Overlaps in Legal Descriptions

16

Richard F. Bales

## I. Introduction

Every real estate attorney and title examiner learns at an early age that legal descriptions whereby the same tract of land is measured and described from two different directions should be avoided. For example, if the Westerly 50 feet of a 100-foot lot is to be conveyed, the remainder should be conveyed as "lot 1, except the Westerly 50 feet." The remainder should not be conveyed as "the Easterly 50 feet of lot 1," as often a surplus or deficiency exists within a tract which would produce, upon division, an overlap (one parcel's boundary being located inside the boundary of the other parcel) or gap (a space between the two parcels conveyed). The division of a tract of land via two metes-and-bounds descriptions where the descriptions start from two different directions can produce similar results.[1]

1. Maley & Thuma, *Practical Observations on Problems in Legal Descriptions,* 43 ILL. B.J. 34 (1954). Gaps or overlaps can also result from the division of irregularly shaped parcels. To avoid this problem, the Northerly, Southerly, etc. portion of an irregularly shaped parcel should be measured at right angles, or radially, to or from the Northerly, Southerly, etc. line of such parcel. If this is not done, a gap or overlap can result, particularly if the width of one portion of the parcel is measured at right angles to a side, while the width of the remainder of the parcel is measured by a line drawn parallel to such side. *See* G. WATTLES, WRITING LEGAL DESCRIPTIONS 4.10–4.12 (1976).

Nonetheless, at least in Illinois, such descriptions do exist. Title insurers are often asked to insure the ownership of these gap and overlap parcels. Unfortunately, there are no Illinois statutes that address this issue. Similarly, there appears to be no directly applicable Illinois case law to guide the title company in issuing insurance or the attorney making the request for insurance. The only cases that even approach this issue are those involving the so-called "apportionment rule" or boundary agreements. The apportionment rule mandates that when a tract of land is subdivided into lots, and title to such lots becomes vested in different persons or entities but the actual aggregate dimensions of such lots are either more or less than the aggregate platted dimensions, the excess or deficiency is to be apportioned among the lots, in direct proportion to the lots' respective areas.[2] The apportionment rule, however, has no bearing where gaps or overlaps occur; it applies only where there is an excess or deficiency *within* the divided tract and not, as here, *between* two divided tracts.

The case law involving boundary agreements sometimes may be applicable to gaps and overlaps, but there are limitations on its applicability. Boundary agreements are discussed in greater detail below.

Though there is no apparent Illinois case or statutory law directly on point,[3] this does not mean that the attorney and title examiner are left totally without redress. Enough principles can be gleaned from Illinois case law and statutes to provide a foundation on which the title insurer can make a well-reasoned decision on a gap or overlap problem that can meet the concerns of the property owners. Each decision will depend on the particular facts of the situation.

## II. Overlapping Legal Descriptions

Consider, then, the problem of the overlap using the following example:

> Lot 1 is purportedly a 100-foot-wide lot. Smith conveys the West 50 feet to A; he subsequently conveys the East 50 feet to B. In reality, though, the lot is 99 feet wide, resulting in a 1-foot overlap.

In this situation, the title insurer who is asked to insure title to either parcel will conclude that the deed to A, being first in time, is the "senior" deed and that A, the first owner, has title to the full 50 feet. Further, the title insurer will conclude

2. Balzer v. Pyles, 183 N.E. 215 (Ill. 1932); Evers v. Watkins, 390 N.E.2d 612 (Ill. App. Ct. 1979); *but see* Westgate v. Ohlmacher, 96 N.E. 518 (Ill. 1911); City of Mt. Carmel v. McClintock, 40 N.E. 829 (Ill. 1895).

3. There is, though, case law in other jurisdictions. R. Skelton, The Legal Elements of Boundaries and Adjacent Properties 210 (1930) cites two cases: *Bryant v. Terry,* 225 S.W. 242 (Ky. 1920) and *State v. Stutzner,* 95 So. 701 (La. 1923). Clark on Surveying and Boundaries 306–07 (J. Grimes ed., 1986) notes a myriad of East Coast decisions. *Also see* J.B. Nielsen, Title and Escrow Claims Guide 368–77 (1996). While not referring to any specific cases, other texts also discuss this problem. *See, e.g.,* G. Wattles, *supra* note 1, at 11.16–11.17; C. Brown, W. Robillard, & D. Wilson, Boundary Control and Legal Principles 179–80 (3d ed. 1986) [hereinafter Boundary Control and Legal Principles].

that the deed to B, being second in time, is the "junior" deed and that B, the junior owner, has title to only the remainder, or 49 feet. The basis for this conclusion is the well-established "doctrine of paramount title."[4] Under this doctrine, a deed to property that the grantor has already conveyed is inoperative to pass any title. In this instance, Smith first conveyed the West 50 feet to A. (Such a deed, being first in time, is termed the "senior" deed.) Since the lot is only 99 feet wide, when Smith conveyed the East 50 feet to B by a second or "junior" deed, he actually conveyed only the East 49 feet, since the overlap parcel—the East 1 foot of the West 50 feet—had already been conveyed to A.

Rarely are overlap problems as easy to solve as the example illustrated above. Usually, the title examiner must go back many years in the chain of title of both tracts of land to the time when one person owned both tracts. He or she must then examine the deeds executed by that person to divide the overall parcel in a manner that created the two tracts. If multiple conveyances created the overlap, the issue becomes even more complicated. Only when all of the deeds causing the overlap have been identified can the examiner apply the doctrine of paramount title to see which deed is senior. Under this doctrine, the ultimate owner in the senior deed's chain of title will be the owner of the overlap parcel.

Some title insurance underwriters feel that, had the grantor known of the possibility of an overlap at the time he made the first conveyance, he would have rewritten the legal description of the two parcels so that the overlap would be eliminated. Because of this, they feel that any overlap should be underwritten by proration or by boundary agreement. However, such an argument is a compromise position that attempts to partially satisfy both owners and is not consistent with the doctrine of paramount title. If, under the above example, Smith first conveyed the West 50 feet to A, it seems inequitable to adjust the boundary by proration or boundary agreement so that A receives less than the 50 feet originally conveyed to him.

When two conveyances are delivered simultaneously, there will be no obvious junior or senior deed, and the overlap must be underwritten in a different fashion. Generally speaking, as between the parties to a deed but subject to the rights of third parties, title to land in Illinois passes on the date of delivery of the deed and not the recording date of the deed.[5] Thus, if the two deeds creating the gap are dated the same day, then regardless of when they were recorded, they should be treated as being simultaneous.[6] In such instances, the gap may be underwritten by proration or by boundary agreement, both of which are discussed below.[7]

4. Weihe v. Lorenz, 98 N.E. 268 (Ill. 1912); 13 Illinois Law and Practice, Conveyances § 29, 101, 111; *see also* Smith v. Grubb, 84 N.E.2d 421 (Ill. 1949).

5. 35 I.L.&P. Vendor & Purchaser § 9; 13 Illinois Law and Practice, Conveyances §§ 34, 36, 44, 88; Creighton v. Elgin, 56 N.E.2d 825 (Ill. 1944); Nelson v. Streu, 156 N.E. 313 (Ill. 1927).

6. Schaeppi v. Glade, 62 N.E. 874 (Ill. 1902).

7. Wattles, *supra* note 1, at 11.16–11.17.

## III. Gaps in Legal Descriptions

Using the following example, consider the problem of the gap:

> Lot 2 is purportedly a 100-foot lot. Jones conveys the West 50 feet to C; he subsequently conveys the East 50 feet to D. However, the lot is 101 feet wide, resulting in a one-foot gap between the two tracts owned by C and D.

Like overlaps, gaps are often created many years prior to being discovered. In determining the eventual owner of the gap parcel, the title examiner must go back in the chain of title of all three parcels (the two portions conveyed to C and D and the gap) to the time when one person owned all three parcels. Because gap parcels are not "of record," the title examiner cannot rely on tax maps or tax assessor records in making his determination as to ownership. Rather, he must examine the first two deeds that divide the tract, and then trace the chain of title of both deeds in order to attempt to determine the current owner of the gap.

It is a simple matter for the title insurer to initially take the position that since the original grantor, or his heirs, owns this gap parcel, a deed from that grantor or his descendants is necessary in order for a title company to insure the gap.[8] But what if the grantor or his heirs cannot easily be located? While some title insurance underwriters may take the position that the gap is unmarketable, there are general principles with respect to title matters by which title to the gap parcel can be underwritten.

There is a general rule of construction that mandates that when there is doubt concerning the meaning of a conveyance, the conveyance should be construed against the grantor and in favor of the grantee.[9] If the gap results from the execution of two noncontemporaneous deeds, this rule can be applied to the example cited above as follows. In the first deed, Jones conveyed the West 50 feet of lot 2 to C. In the second deed, Jones conveyed the East 50 feet of lot 2 to D. At this point, an ambiguity arises: did Jones also intend to convey the "gap"—the adjacent West 1 foot of the East 51 feet of lot 2—to D, or did Jones intend to retain this portion of the lot? Using the rule of construction set forth above, it is possible to construe the language in the deed in favor of the grantee, D, to the detriment of the grantor, Jones. Such an interpretation would result in the entire gap being vested in favor of the second grantee, D.

Unfortunately, the issue is not that simple. There is another rule of construction that states that the primary rule in construing a deed is to ascertain the intent of the grantor and, when possible, to give effect to such intent.[10] Clearly, using the examples above, Jones intended to give the West 50 feet of Lot 2 to C. One might maintain, also, that it is clear that he intended to give the East 50 feet

8. Boundary Control and Legal Principles, *supra* note 3, at 179–80.

9. Williams v. Swango, 7 N.E.2d 306 (Ill. 1937); 13 Illinois Law and Practice, Conveyances § 76; Grigsby, *Descriptions and Plats,* 39 Ill. B.J. 439 (1951).

10. Bibo v. Bibo, 74 N.E.2d 808 (Ill. 1947); Smith v. Grubb, 84 N.E.2d 421 (Ill. 1949).

of Lot 2 to D. But, did Jones intend to keep the gap to himself or did he intend to relinquish all interest in the remaining property at the time he made the second conveyance? One might very well argue the latter, that it is doubtful that Jones intended to retain ownership of the gap, especially if there were no "reservation of gap" language in the deed or if the gap occurred many years prior to being discovered. Particularly where Jones or his heirs no longer own any other property contiguous to lots owned by C and D, there seems no reason to believe Jones or his heirs would claim any remaining interest in Lot 2. Accordingly, the primary rule in construing a deed—to ascertain and give effect to the grantor's intent—like the first general rule, favors D, the second grantee. Indeed, it is because of these two rules of construction that many people feel that a gap parcel vests in this second grantee.[11]

But are these two rules sound in their application to this problem? What if the deeds were executed simultaneously? Furthermore, is it truly reasonable to state, concerning the above example, that Jones intended to give all of the gap parcel to D, the second grantee? Might not one maintain, just as easily, that Jones never intended to create a gap at all?[12] That is, it seems reasonable to conclude that had Jones known of the possibility of a gap at the time he made the first conveyance—that is, had he known that Lot 2 was really 101 feet wide—he would have rewritten the legal descriptions of the two parcels so that the gap would be eliminated.

Because either possibility has merit and the other principles discussed above do not address them, several rules for underwriting "gaps" have been developed, each of which is discussed separately.

These methods do not dispute the possibility that title to the gap remains vested in the original grantor. However, after considering factors such as the size and age of the gap,[13] some title insurers may consider the underwriting of title to this gap parcel by proration or boundary agreement to be an acceptable risk.

---

11. *See, e.g.*, CLARK, *supra* note 3, at 290–91 ("The court held that [the gap] was not to be apportioned between A and B but should go entirely to B. This decision was based on the theory that the tract conveyed to A, being definite and certain and being first conveyed, left the balance of the land in the grantor and that he subsequently sold all of said balance to B. If all of the tract had been conveyed at the same time to the two grantees and there were no monuments marking the line between the two, it is more than likely that the court would have apportioned the excess to the two grantees."). *See also* Adams v. Wilson, 34 So. 831 (Ala. 1903).

12. *See Brenneman v. Dillon,* 296 Ill. 140 at 148, 129 N.E. 564 (1921), where the court notes that while deeds are to be construed in favor of the grantee, "conflicting descriptions should be reconciled, if possible, and the construction adopted which best comports with the intention of the parties." *See also* 13 Illinois Law and Practice, Conveyances §§ 76, 78.

13. The age of the gap, as noted earlier, is determined by going back to that point in time when one entity owned the gap parcel and the adjoining parcels. The date of the second of the first two deeds that divides this tract is the "birth date" of the gap.

## IV. Proration of a Gap Parcel

A proration of the gap is merely an extension of the theory that the owner of the entire parcel never intended to keep any portion of the parcel he divided but rather meant to convey the entire property on a proportionate basis in the two deeds.[14] Indeed, this is possibly the only reasonable interpretation of the "intention of the parties" doctrine discussed earlier. Thus, if Able, believing he owned a 300-foot-wide lot, conveyed the West 100 feet to E and the East 200 feet to F, any gap would be prorated, one-third to E and two-thirds to F. Or, if Baker, thinking he owned five acres of land, conveyed two acres to G and three acres to H, any resultant gap would be prorated, two-fifths to G and three-fifths to H. The proration method is probably the least controversial means of underwriting the gap problem. Since both parties are assigned some portion of the gap (i.e., one lot owner is not assigned the total width of the gap), the risk to the title insurer that one of the parties will file suit, claiming title to the entire gap, is probably minimal. Nonetheless, note that while this proration method is an acceptable means of underwriting the gap problem, some title insurers may not wish to utilize it unless a deed from the adjoining parcel owner is also obtained and recorded.

## V. Boundary Agreements to Eliminate a Gap

A gap can also be insured by a boundary agreement. One can establish boundary lines by agreement in only two situations: either when a boundary line is in dispute or when the boundary line is unascertained.[15] Normally, the statute of frauds requires that a change in boundary lines be accomplished by one or more conveyances from one party to another. However, in determining the location of an unascertained or disputed boundary line, applicable Illinois decisions take the position that the boundary agreement merely defines the actual boundary line; as a result, no transfer of land from one party to another occurs.[16]

Since a boundary agreement can be used only for establishing a disputed or unascertained line, one cannot establish a boundary by agreement if the landowners know the true location of the boundary line, because the required elements of uncertainty or dispute needed to except the agreement from the statute of frauds are not present. Similarly, if the landowners know the true location of the

14. G. Wattles, *supra* note 1, at 11.16–11.17.

15. I.L.&P. Boundaries §§ 31 et seq. Loverkamp v. Loverkamp, 45 N.E.2d 871 (Ill. 1943). Note how the court in *Skinner v. Francisco,* 88 N.E.2d 867, 870 (Ill. 1949) defines "uncertainty": "A deed or other written contract is not void for uncertainty in a description of the land conveyed if, from the words employed, the description can be made certain by extraneous evidence of physical conditions, measurement or monuments referred to in the deed."

16. J. Backman & D. Thomas, A Practical Guide to Disputes between Adjoining Landowners–Easements 8–12, 8–15 (1991); *see also* Nitterauer v. Pulley, 82 N.E.2d 643 (Ill. 1949).

boundary line, they cannot by agreement change the location. As the court in *Hartzler v. Uftring* noted:

> When an unascertained or disputed boundary is established . . . it will be binding on the parties to the agreement and their privies in estate. The boundary, once established, will control the parties' deeds notwithstanding the statute of frauds. The principle upon which this conclusion is reached is that the effect of the agreement is not to pass real estate from one party to another but simply to define the boundary line to which their respective deeds extend. The requirement that the boundary be unascertained or in dispute is therefore a necessary prerequisite to any agreement, since if the location of the true boundary line is known to the owners they cannot transfer the land from one to the other by an agreement changing such location.[17]

Consistent with this principle, the Illinois Supreme Court case has held that if the intention of the parties is to ascertain the true boundary line, and in so doing an erroneous line is agreed upon either by accident or mistake, the boundary agreement is not binding and the line will not be established by virtue of the agreement.[18] The reason for this is that in this instance, there is a failure to find the true line and not an agreement to adopt an unascertained or disputed line.[19]

## VI. Parol Boundary Agreement versus Written Boundary Agreement

Illinois case law indicates that if a boundary line is established only by parol agreement, there must be subsequent possession consistent with the agreement.[20] This possession, however, need not be for the statutory period of twenty years required to establish title by adverse possession.[21] On the other hand, if the boundary agreement is in writing, possession consistent with the agreement is not required. As the court in *Purtle v. Bell* stated:

> Where there is a dispute between adjoining land owners as to the true boundary line between their respective lands, or if the line between them is unascertained, they have the right, in law, to establish a boundary line by agreement, either in writing or by parol, but when by parol, possession must be taken in pursuance thereof, and the line so agreed upon will be binding upon the parties and their privies in estate.[22]

17. Hartzler v. Uftring, 114 Ill. App. 3d 427, 430, 450 N.E.2d 1208 (1983).
18. Sonnemann v. Mertz, 77 N.E. 550 (Ill. 1906).
19. Ginther v. Duginger, 129 N.E.2d 147 (Ill. 1955).
20. McLeod v. Lambdin, 174 N.E.2d 869 (Ill. 1961).
21. Horn v. Thompson, 58 N.E.2d 896 (Ill. 1945).
22. Purtle v. Bell, 225 Ill. 523, 525, 80 N. E. 350 (1907)

Note, however, that most title insurers would probably not insure title based on a boundary line established only by parol agreement.

## VII. The Boundary Agreement and the Permanent Survey Act

Many people believe that a boundary agreement can be utilized even when the boundary is not in dispute or is unascertained. As evidence of this premise they point to § 1 of the Permanent Survey Act, codified as 765 Ill. Comp. Stat. 215/1, which provides:

> That whenever the owner or owners of adjacent tracts of land shall desire to establish permanently the lines and corners thereof between them, he, she or they may enter [into] a written agreement to employ and abide by the survey of some surveyor, and after said survey is completed, a plat thereof with a description of all corners and lines plainly marked thereon, together with the written agreement of the parties, shall be recorded in the recorder's office of the county where the lands are situated; and the lines and corners of said survey so made and recorded, shall be binding upon the parties entering into said agreement, their heirs, successors and assigns, and shall never be changed.[23]

Reliance on this statute is erroneous for several reasons. First of all, statutes are to be construed in harmony with the existing law as part of a general and uniform system of jurisprudence. The Permanent Survey Act must, therefore, be interpreted consistently with Illinois case law.[24] As already mentioned, Illinois case law makes it clear that the two elements—a boundary in dispute or an unascertained boundary—are necessary before a boundary agreement can be utilized.

Second, when construing a statute in order to determine the intent of the General Assembly, the statute should be construed as a whole so that the legislative intent is gleaned from the entire statute, rather than from any one part of it. The statute should be construed, if possible, so that no word, clause, or sentence is rendered superfluous or meaningless.[25] In this regard, the wording of § 2 of the Permanent Survey Act, codified as 765 Ill. Comp. Stat. 215/2 is relevant. It provides:

> Whenever one or more proprietors of lands in this state, the corners and boundaries of whose lands *are lost, destroyed, or are in dispute,* or who are desirous of having said corners and boundaries permanently re-established, *and who will not enter into an agreement as provided by section one of this*

23. *See generally* 5 Illinois Law and Practice, Boundaries § 29, et seq.

24. 34 Illinois Law and Practice, Statutes § 54, et seq. § 130; 82 C.J.S. *Statutes* § 362; People v. Clark, 108 N.E. 994 (Ill. 1915).

25. 34 Illinois Law and Practice, Statutes §§ 64, et seq. §§ 123 et seq.; Ill. Bell Tel. Co. v. Ames, 4 N.E.2d 494 (Ill. 1936); Peacock v. Judges Ret. Sys. of Ill., 140 N.E.2d 684 (Ill. 1957).

> *act, it shall be lawful* . . . [to] make application to the circuit court of the county in which said lands are situated, for the appointment of a commission of surveyors to make survey of and to permanently establish said corners and boundaries. . . . (emphasis added).

It is evident that § 2 of the Permanent Survey Act exists as an alternative to § 1—that is, if the landowners are unwilling to enter into a boundary agreement, pursuant to § 1, then other relief is available to the parties, pursuant to § 2.[26] If these two sections of the Act are read and interpreted together, it seems clear that the establishment of boundary lines pursuant to a boundary agreement as noted in § 1 of the Act is predicated on boundary lines that are in dispute or unascertained pursuant to § 2 of the Act.

Finally, in cases of ambiguity, a statute should be construed with reference to the common law that was in force at the time of the statute's passage.[27] The Permanent Survey Act was enacted in 1901. The Illinois Supreme Court in *Yates v. Shaw* made the following comments concerning boundary agreements:

> Lines and corners that are supposed to be fixed and established by one surveyor are overturned or left in doubt by another, at a subsequent period. *In all matters of uncertainty and dispute,* the parties may, without doubt, compromise and end the dispute. And they may as certainly fix by agreement, the boundary lines separating their lands, as other disputes.[28]

In the *Yates* case, forty-one years before the passage of the permanent Survey Act, the Illinois Supreme Court defined those instances in which a boundary agreement may be utilized. It is clear that if this statute were construed consistently with this case, the elements of a boundary in dispute or an unascertained boundary would be necessary elements before a boundary agreement could be utilized to establish title to a gap or overlap.[29]

## VIII. An Exception to the Rules Regarding Boundary Agreements

There is one potential exception to the requirement that there be either a boundary in dispute or an unascertained boundary before a boundary agreement can be used to establish title. A survey can indicate not only the boundary line, or ownership line, between two parties, but also the record title line relative to this

26. Kelch v. Izard, 590 N.E.2d 1050 (Ill. App. Ct. 1992).

27. 34 Illinois Law and Practice, Statutes § 71; 82 C.J.S. *Statutes* § 363; Livingston v. Meyers, 129 N.E.2d 12 (Ill. 1955).

28. 24 Ill. 368, 369 (1860) (emphasis added).

29. Corrington v. Pierce, 28 Ill. App. 211 (1888), McNamara v. Seaton, 82 Ill. 498 (1876).

boundary or ownership line.[30] If the survey shows gaps and/or overlaps between an ownership line and a record title line, a boundary agreement may be drafted that calls for the exchange of quitclaim deeds between the owners. By utilizing such an agreement and then issuing quitclaim deeds consistent with the agreement, the gaps and/or overlaps can be eliminated, thus reconciling the ownership lines and the boundary line.

The use of a boundary agreement in this instance is perfectly proper, even though the boundary line is not disputed and is ascertained, because the agreement is not really a boundary agreement at all. A true boundary agreement is used to establish a boundary line. Here, the location of the boundary is not at issue as it is already known and established. Instead, the agreement is merely a vehicle used, together with one or more conveyances, to conform an ownership line to a record title line.

## IX. Recently Created Gaps

If the gap were created many years prior to its discovery, it is reasonable for a title insurer to rely on a proration of the gap or a boundary agreement as a means of insuring the gap without requiring a deed from the grantor or his heirs. This underwriting procedure stems from a doctrine of deed construction that allows one to consider the surrounding or attendant circumstances if such circumstances indicate the practical construction of the deed by the parties to the deed.[31] Thus, when a gap is created years ago and the gap is not shown on any tax maps of the property and the grantor is not exercising any dominion or control over the gap, the risk to most title insurers of not requiring a deed from the grantor or his descendants is probably minimal. Consequently, underwriting by proration or boundary agreement is no doubt acceptable to many insurers.

However, if the gap is of relatively recent vintage, ownership of this parcel should not be underwritten by proration or by boundary agreement. Instead, one or more deeds from the owner of the gap parcel (i.e., the original grantor) should be executed and recorded. The reason for this, again, goes to the issue of intent. Is it reasonable under these circumstances to rely on the fiction that this grantor who recently conveyed the adjoining property also intended to convey the gap,

30. An example would be two adjacent subdivisions wherein the supposed common line contained in the legal descriptions of the underlying deeds of the two subdivisions (i.e., the ownership line) are not coincident with the supposed common line contained in the legal descriptions of the recorded plats (i.e., the record title line). Another example would be two tracts of land wherein the owner of one tract has acquired a portion of the adjoining tract through adverse possession. For a further discussion of the latter example, see note 33 and text discussion *infra.*

31. 113 Illinois Law and Practice, Conveyances § 86; Henry v. Metz, 46 N.E.2d 945 (Ill. 1942); Hollenbeck v. Hollenbeck, 83 N.E. 926 (Ill. 1908). As the court stated in *Law v. Kane,* 52 N.E.2d 212, 215 (1944): "Where the deed contains ambiguous language, a reasonable interpretation by the parties, themselves, found in their contemporaneous or subsequent acts and declarations, will generally be adopted."

when in reality the grantor still owns it? Is it reasonable to second-guess this grantor as to how the gap should be divided between the two adjacent landowners, if divided at all?

Thus, if a title insurer is asked to underwrite the gap, insuring title in one or both of these adjacent landowners, and the deeds creating the gap were recorded fairly recently, at least some attempt should be made to obtain a deed or deeds from the grantor of this gap parcel. In this manner, the grantor can decide the ultimate disposition of the gap. If this is not possible, the title insurer should consider the reasons this is so. If, for example, the grantor is available but uncooperative, the title insurer may decline to underwrite the gap, even on a "risk basis,"[32] pursuant to proration or a boundary agreement. Instead, the attorney may have to institute a quiet title action on behalf of his or her client.

## X. Surveys and Adverse Possession

A survey should be obtained prior to insuring the gap, preferably a land title survey.[33] The reason for this is that the gap may be underwritten solely by proration or boundary agreement only if there is no evidence of adverse occupation onto the gap parcel.[34] Such reasoning is obvious; a gap should not, e.g., be prorated between grantees A and B when the survey discloses adverse possession by A onto what would be B's share of the gap (e.g., A's fence is located on a portion of the gap that would, by proration, be assigned to B). Under such circumstances, one or more quitclaim deeds of the occupied property should be prepared and recorded so that the ownership lines, as altered by the adverse possession, conform to the prorated boundary line. If this is not possible, an exception for the possible adverse interests of A would have to be raised in Schedule B of any title policy issued covering B's parcel.

If, on the other hand, the survey showed adverse possession by A onto B's portion of the gap and the title insurer was insuring A's portion of the gap, an exception would be added for the possible encroachment by A's improvement onto B's adjoining property. In some instances, an encroachment endorsement

32. Generally speaking, the role of a title insurer is risk elimination and not risk assumption. Nonetheless, there may be instances when a title insurance company chooses to assume a known title defect. As noted in MICHAEL J. ROONEY & JOHN A. THORNHILL JR., ILLINOIS BOUNDARY LAW AND ADJOINING LANDOWNER DISPUTES (ADDENDUM, 1988), at 2–4, a title company may decide to assume such a defect provided a cash deposit is made with the company so that it may be reimbursed for any losses paid. Barring such an indemnity, however, it is doubtful that a title insurance company would choose to underwrite a gap created by recently recorded deeds without the recording of a deed or deeds from the grantor of the gap parcel.

33. Miely, *ALTA/ACSM Survey Standards,* 37 REAL PROPERTY 1 (1991); *see also Illinois Land Survey Standards,* 57 ILL. B.J. 327 (1968).

34. *See generally* 735 ILL. COMP. STAT. 5/13–101 et seq.; 1 Illinois Law and Practice Adverse Possession §§ 1 et seq.; Joiner v. Janssen, 421 N.E.2d 170 (Ill. 1981).

insuring A against loss by reason of being unable to maintain the encroachment might be available, depending on the nature of the encroachment.

If, in this same example, the title company were insuring B's share of the gap, such an exception would be worded as an encroachment by A's improvements onto the land in question being insured for B's benefit. From the title insurer's perspective, there is, of course, no endorsement that can be given on B's owner's policy for such an encroachment. To do so would be to invite the insured, B, to make a claim the moment the policy was issued.[35]

At first, one might think that the insuring of a gap via a true boundary agreement would be permissible, even if there is evidence of adverse occupation of the gap parcel. It is feasible that one might rationalize that the execution of a boundary agreement whereby the parties agree to a boundary line would negate any adverse possession of the gap.

Such thinking is incorrect. First of all, it is possible that title to the occupied portion of the gap might already have passed to an adjoining owner by virtue of adverse possession.[36] Thus, the execution of any boundary agreement could not negate the vested consequences of this adverse occupation, as it would not be executed by the true owner of the occupied parcel. Second, title to the gap could not be conveyed by a boundary agreement because a boundary agreement is an instrument of agreement, to be used to establish an unascertained or disputed boundary line, and not an instrument of conveyance, to be used to convey real estate from one party to another.[37]

## XI. Real Estate Taxes

Finally, the title insurer, when insuring the gap, must consider the raising of an exception for "back taxes" or "omitted taxes" on any title insurance commitment or policy.[38] The rationale for this stems from the provisions of 35 Ill. Comp. Stat. 200/9–260 et seq.,[39] which provides for the assessment of real estate tax arrearages against property which had, in previous years, been omitted from the tax rolls. This exception can be waived if the title insurer is furnished evidence that no such taxes will be assessed (which may be difficult to obtain) or purely on a risk basis.

This issue may, practically speaking, be theoretical at best. Most tax assessment legal descriptions are vague and not complete; they could encompass any

35. Attorney's Guide to Title Insurance 7–17 (Ill. Inst. for CLE, 1980).

36. Even though title vests in the adverse occupier at the time all of the elements of adverse possession are satisfied, it is doubtful that a title insurance company would insure title accordingly unless the adverse possession was confirmed by a court order.

37. *See* Jones v. Scott, 145 N.E. 378 (Ill. 1924); Hartzler v. Uftring, 450 N.E.2d 1208 (Ill. App. Ct. 1983).

38. Such an exception might read as follows: "Possible back taxes which may be assessed against the property as a result of the non-assessment of the land in prior year(s)."

39. *See also* 35 Ill. Comp. Stat. 200/9–270; 200/16–50; 200/16–135; 200/14–45.

gap in record title. Thus, any Supervisor of Assessments would probably take the position that the gap parcel would have an assessment, albeit an incorrect one, and thus, there would be no omitted taxes.

Nonetheless, there is statutory authority for the assessment of a gap parcel. This is noted in said 35 Ill. Comp. Stat. 200/9–260:

> [T]he county assessor shall have power . . . to assess properties which may have been omitted from assessments for the current year or during any year or years for which the property was liable to be taxed, and for which the tax has not been paid. . . .

## XII. Conclusion

Gap and overlap problems are not common, but the title insurer and practitioner will encounter them from time to time. The above guidelines should be more than sufficient to aid the title company and real estate attorney in analyzing and underwriting the problem to the satisfaction of both the insurer and its insured.

# Center of Section— Center of Controversy?

17

Richard F. Bales

At first glance, one might think that locating the center of a section is simple enough. However, because of conflicting law and instructions, this has been a matter of controversy for more than one hundred years. The purpose of this article, then, is twofold. One, it will discuss the history of the "center of section" problem, putting all facets into a historical perspective. Two, it will attempt to outline both a legal and practical solution to the problem. While some emphasis will be on Illinois law, surveyors in all jurisdictions should find the issues discussed to be of benefit.

The scenario starts in the early 1800s—specifically, the Land Act of February 11, 1805 [hereinafter the 1805 Act]:

> Sec. 2. That the boundaries and contents of the several sections, half sections, and quarter sections, of the public lands of the United States, shall be ascertained in conformity with the following principles, any act or acts to the contrary notwithstanding:
>
> 1. All the corners marked in the surveys, returned by the surveyor general, or by the surveyor of the land south of the State of Tennessee, respectively, shall be established as the proper corners of sections, or subdivisions of sections, which they were intended to designate; and the corners of half and quarter sections, not marked on the said surveys, shall be placed, as nearly as possible, equidistant from those two corners which stand on the same line.

> 2. The boundary lines, actually run and marked in the surveys returned by the surveyor general, or by the surveyor of the land south of the State of Tennessee, respectively, shall be established as the proper boundary lines of the sections or subdivisions for which they were intended; and the length of such lines, as returned by either of the surveyors aforesaid, shall be held and considered as the true length thereof. And the boundary lines which shall not have been actually run and marked as aforesaid shall be ascertained by running straight lines from the established corners to the opposite corresponding corners. . . . [1]

The 1805 Act stands for the general proposition, often cited, that surveyors must locate the center of a section at the intersection of straight lines connecting opposite quarter section corners. Some writers feel that the 1805 Act contains inconsistencies—discrepancies that may have been responsible for the subsequent different methods of determining the center of a section. Indeed, the Bureau of Land Management states as much in its book, SUBDIVISION OF SECTIONS, when it maintains that the first section of the 1805 Act is superfluous when compared to section 2, paragraphs one and two, quoted above.[2] However, Curtis M. Brown, Walter G. Robillard, and Donald A. Wilson, in their book, BOUNDARY CONTROL AND LEGAL PRINCIPLES, state that section 1 merely applied to lands surveyed under the Act of May 18, 1796.[3] They state further, though, that the first paragraph of section 2 is in conflict with the second paragraph of this section. They maintain that the second paragraph gives the general rule for determining the center of section, but that the first paragraph, which states that the "corners of half and quarter sections, not marked on the said surveys, shall be placed, as nearly as possible, equidistant from those two corners which stand on the same line" is in conflict with this rule.[4] One might argue, however, that the second paragraph gives the method for locating the center of section (or center quarter corner), while the first paragraph not inconsistently determines the location of the North, South, East, and West quarter corners.

Notwithstanding the 1805 Act, Justin Butterfield, Commissioner of the General Land Office, issued special instructions from 1849 to 1851, which directed that the center of the section be located at the midpoint of the line connecting the East quarter section corner and the West quarter section corner.[5]

Similar instructions were in use from 1856 to 1883; they were issued to deputy surveyors for the District of Illinois and Missouri: "run a true line from [the East

1. Joe D. Webber, EARLY PUBLIC LAND SURVEYS IN THE NORTHWEST TERRITORY AND PROCEDURES FOR THE RETRACEMENT OF ORIGINAL GOVERNMENT SURVEYS IN ILLINOIS 234–35 (1981); *see also* CURTIS M. BROWN, WALTER G. ROBILLARD, & DONALD A. WILSON, BOUNDARY CONTROL AND LEGAL PRINCIPLES 256–57 (3d ed. 1986); U.S. BUREAU LAND MGMT., SUBDIVISION OF SECTIONS 16–19 (1980); JOE D. WEBBER, FEDERAL INSTRUCTIONS FOR SURVEYS OF THE PUBLIC LANDS FROM 1785–1843, at 57–60 (1975).

2. SUBDIVISION OF SECTIONS, *supra* note 1, at 17.

3. BROWN ET AL., *supra* note 1, at 256.

4. BROWN ET AL., *supra* note 1, at 257.

5. BROWN ET AL., *supra* note 1, at 318–19. This procedure of locating the center of a section is sometimes called the "East–West split."

quarter section corner] to [the West quarter section corner], and at equidistance between them establish [the center of the section]."[6]

It is important to note that, unlike the 1805 Act, both sets of instructions were merely advisory opinions. As the 1856 instructions stated: "It is not intended, by what is here recommended for . . . subdividing Sections, to give any *positive directions* to county Surveyors. This office has no control over them whatever. . . ."[7]

Clearly, both sets of instructions were contrary to the 1805 Act. Confused Illinois surveyors sought an opinion from Abraham Lincoln, a prominent surveyor and lawyer. He wrote the following response just two years before he assumed the presidency:

> The eleventh section of the Act of Congress, approved February 11, 1805, prescribing rules for the subdivision of sections of land within the United States system of survey, standing unrepealed, in my opinion, is binding on the respective purchasers of different parts of the same sections and furnishes the true rule for surveyors in establishing lines between them. That law, being in force at the time that each became a purchaser, becomes a condition of the purchase. And by that law, I think the true rule for dividing into quarters, any interior section, or section which is not fractional is to run straight lines through the section from the opposite quarter-section corners, fixing the point where the straight lines cross, or intersect each other, as the middle or center of the section.
>
> Nearly, perhaps quite, all the original survey's [sic] are to some extent erroneous, and in some of the sections, quite so. In each of the latter, it is obvious that a more equitable mode of subdivision than the above might be adopted; but as error is infinitely various, perhaps no single better rule can be prescribed.
>
> At all events, I think the above has been prescribed by a competent authority.
>
> Springfield, Illinois
>
> January 6, 1859
>
> A. Lincoln[8]

Despite Lincoln's opinion, many surveyors in Illinois and other states continued to locate the center of section in accordance with the 1856 instructions. Some surveyors, in fact, even went so far as to locate the center not at the midpoint of that line connecting the East and West quarter section corners, but rather at the midpoint of that line connecting the North and South quarter section corners! The

---

6. Webber, *supra* note 1, at 338, 345–48. These instructions, in use from 1856 to 1883, are hereafter termed the "1856 instructions."

7. Webber, *supra* note 1, at 331.

8. T.S. Madson II & Louis N.A. Seemann, Fading Footsteps (or, Retracement and the Land Surveyor) 79, 115–16 (1980).

surveyor of today, then, levels his total station on the horns of a dilemma: when he is surveying property whose legal description is dependent on a center of section located contrary to the 1805 Act, does he follow this precedent and perform his survey accordingly, but possibly in violation of federal law, or does he survey according to the 1805 Act and possibly disturb the property rights of adjacent landowners?

The answer, it seems, may be jurisdictional in nature. As Brown, Robillard, and Wilson point out:

> Once land passes from the federal government to private parties within a state, the jurisdiction over court trials passes from the federal courts to the state courts. While most state courts recognize federal court rules as applicable in resurveys, this is not universally true. For example, the Missouri state court accepts that state's statute regarding restoration of a lost section corner, which is quite different from the requirements of federal courts; and in California courts topography calls have much greater force than that given in federal courts.[9]

They also give a general principle:

> When resurveying or dividing federally created sections of land, federal rules for resurvey are followed (1) when the final court of adjudication resides in a federal court or (2) when the state court has approved federal court rules. If the state court's rules for resurvey or for dividing sections differ from those of the federal court and if the state court has final jurisdiction, the state court's opinions may be followed.[10]

The surveyor, then, must first determine whether his resurvey is governed by federal law or state law. What follows is a general discussion of both.

Any analysis of federal law must be on a two-tiered approach. First, it must define a retracement survey, and second, it must address the rights of third parties who acquire an interest in the property after the initial survey but prior to the resurvey.

In defining a retracement survey, the U.S. Supreme Court, in the landmark case, *Cragin v. Powell,* 128 U.S. 691 (1888), stated: "A resurvey, properly considered, is but a retracing, with a view to determine and establish lines and boundaries of an original survey. . . .[11]

Federal law regarding retracements thus seems to echo the classic surveying texts. The following, for example, appears in BOUNDARY CONTROL AND LEGAL PRINCIPLES: "The responsibility of the retracement surveyor is to follow the footsteps of the

---

9. BROWN ET AL., *supra* note 1, at 291 and also 325.

10. BROWN ET AL., *supra* note 1, at 291.

11. Cragin v. Powell, 128 U.S. at 698 (quoting U.S. BUREAU OF LAND MANAGEMENT, MANUAL OF INSTRUCTIONS FOR THE SURVEY OF THE PUBLIC LANDS OF THE UNITED STATES 146 (1973) [hereinafter 1973 Manual]).

original surveyor as nearly as possible."[12] Or, consider the following passage from A Treatise on the Law of Surveying and Boundaries: "The cardinal principle guiding a surveyor who is running the lines of a previous survey is to follow in the footsteps of the previous surveyor."[13]

The second federal tier is merely a corollary of the first; it deals with the rights of third parties who may acquire an interest in the resurveyed property. These rights are set forth in the Act of March 3, 1909, 43 U.S.C. § 772: "That no such resurvey or retracement shall be so executed as to impair the bona fide rights or claims of any claimant, entryman, or owner of lands affected by such resurvey or retracement."[14]

It is clear, then, that under federal law, a resurvey of lands originally surveyed under the 1856 (or similar) instructions would have to be similarly resurveyed so as not to disturb the rights of bona fide claimants.[15] This is the case, even though such a resurvey might monument to the "wrong" center of section, i.e., a center that is located contrary to the 1805 Act. Note that these claimants are protected, regardless of whether a boundary agreement exists or whether said claimants have occupied the land for a certain requisite time period. These factors may be necessary under such doctrines as "practical location of boundaries," "boundaries by acquiescence," or "adverse possession"; they are not requirements here.

What, then, about retracement surveys of "center of section" lands under state law? First of all, note that Illinois case law is replete with cases that mime the previously stated "following in the footsteps" philosophy of the retracement surveyor.[16] The author could find no Illinois cases, though, that address the rights of bona fide claimants vis-à-vis these surveys.[17] However, Illinois statutory law may be the key. 55 Ill. Comp. Stat. 125/5 (1992), formerly Ill. Rev. Stat. 1991, ch. 133, sec. 5 states: "All surveys shall be made in accordance with the original surveys and the laws of the United States governing surveys."[18] Clearly, this statute parallels the rule noted as paragraph 2 in the principle set forth in Boundary Control and Legal Principles, quoted *supra:* "When resurveying or dividing federally created sections of land, federal rules for resurvey are followed . . . when the state court has approved federal court rules."[19]

---

12. Brown et al., *supra* note 1, at 324.

13. John S. Grimes, A Treatise on the Law of Surveying and Boundaries 339, 366 (4th ed. 1976).

14. 1973 Manual, *supra* note 11, at 147.

15. Grimes, *supra* note 13, citing a 1955 California case, seems to make the same point when he says: "A survey will never affect titles taken under a prior survey." Grimes, Treatise on the Law of Surveying 340.

16. For example, McClintock v. Rogers, 11 Ill. 279 (1849); Bauer v. Gottmanhouser, 65 Ill. 499 (1872); *see also* Michael J. Rooney & John A. Thornhill Jr., Illinois Boundary Law and Adjoining Landowner Disputes (addendum, 1988); 5 Illinois Law and Practice, *Boundaries* 439–79 (1953).

17. *See, though*, Westgate v. Ohlmacher, 251 Ill. 538, 96 N.E. 518 (1911). In *Westgate*, the court, while referring to Justice Cooley of Michigan and his well-known *Diehl v. Zanger* decision, stated that "the courts ought not to disturb boundary lines between lot owners which have been acquiesced in for years and upon which the lot owners have erected improvements . . . ." 251 Ill. at 541.

18. *See also* 55 Ill. Comp. Stat. 125/10 (1992), formerly Ill. Rev. Stat. 1991, ch. 133, sec. 10.

19. Brown et al., *supra* note 1, at 291.

Thus, Illinois law, like federal law, recognizes the rights of bona fide claimants. A surveyor doing a retracement survey of Illinois lands originally surveyed pursuant to the 1856 instructions would, under Illinois law, have to similarly resurvey said lands.

There is, furthermore, one other argument; while the following reasoning applies to Illinois lands, it may be applicable in other jurisdictions as well.

Some surveyors argue that the early surveyors who surveyed according to the instructions of the mid-1800s did so without authority as these instructions were contrary to the 1805 Act. These surveyors reason that any subsequent survey must be performed in conformity with the 1805 Act, regardless of the adverse effects such a survey might have on the surrounding area.[20]

Such thinking may be incorrect. Consider the following analysis of the legal and theoretical aspects of the "center of section" problem.

The 1849–1851 instructions were issued by the Commissioner of the General Land Office. The 1859–1883 instructions came from the Surveyor General of the General Land Office. This office, the predecessor of the Bureau of Land Management, was a federal agency. *Black's Law Dictionary* defines "agent" as "one who represents and acts for another under the . . . relation of agency."[21] Clearly, these men were agents of the federal government.

The instructions emanating from these men were contrary to the 1805 Act. This act was federal law. Such erroneous instructions, coming from federal agents, were thus unauthorized.[22] However, it is a basic tenet of agency law that an affirmation, or ratification, of an unauthorized act can be inferred from a failure to repudiate said act.[23]

These instructions were in existence for more than twenty-five years. Concerned surveyors met in Springfield, Illinois, to discuss them. No less a man than Abraham Lincoln commented on them. It strains the limits of credibility to maintain that the federal government did not know of these instructions. Indeed, it seems reasonable to assume that the federal government was aware of the instructions but failed to repudiate them. Since countless surveyors performed surveys in reliance on these instructions, the federal government, under agency law, may not be estopped, or prevented, from denying that it had ratified or sanctioned these instructions.[24]

20. For an excellent discussion of this topic, see Boundary Control and Legal Principles 374–76.

21. Black's Law Dictionary 59 (5th ed. 1979).

22. It stands to reason that these instructions were either authorized or unauthorized. But, if they were authorized, there would not be this problem now, more than one hundred years later!

23. Restatement (Second) of Agency § 4.01 (1958).

24. Restatement (Second) of Agency § 2.05 Note that the estoppel referred to here stems from an agency relationship. It is slightly different from the concept of establishing a boundary line by "equitable estoppel." Regarding the latter, see Grimes, *supra* note 13, at 639–43; Illinois Law and Practice, *supra* note 16, at 463–64.

Authors of surveying texts have reached similar conclusions. For example, Joe D. Webber writes in the aforementioned Early Public Land Surveys in the Northwest Territory:

> If existing evidence indicates the lines and corners conform to their establishment by the 1856 instructions, they should not be changed to conform to later instructions. Because these instructions were the retracement rules in force at the time, corners and lines established under their auspicious [sic] would be valid locations, particularly if acquiescence by adjacent land owners or the public could be shown.[25]

On the other hand, some people argue that under federal agency law, there is no such estoppel; i.e., despite the erroneous acts of a federal agency, the federal government is *not* estopped from later denying that it had ratified these 1856 instructions. While this may be true, the preceding analysis is nonetheless valid. Note that 55 Ill. Comp. Stat. 125/5 (1992), formerly Ill. Rev. Stat. 1991, ch. 133, sec. 5, states that "all surveys shall be made in accordance with the original surveys and the laws of the United States governing surveys." Does not a retracement survey prepared pursuant to the 1856 instructions of lands that were originally surveyed in accordance with these instructions meet the criteria of this statute? After all, the retracement survey was made "in accordance with the original surveys"—i.e., since the lands were originally surveyed pursuant to the 1856 instructions, the subsequent retracement survey was similarly prepared. And, this survey was made in accordance with "the laws of the United States governing surveys." That is, although the survey was not prepared pursuant to the 1805 Act, but rather, in accordance with the 1856 instructions, which were admittedly advisory opinions only, the retracement survey nonetheless was prepared pursuant to federal law in that it recognizes the rights of third-party claimants, as set forth in the previously cited federal statute, 43 U.S.C. § 772.

The previous pages approach the "center of section" problem from a historical and legal perspective. For the surveyor in the field, there are, of course, practical considerations as well. A procedure for recognizing and handling a "center of section" problem is outlined below in eleven steps; the procedure was developed by several Illinois land surveyors.

1. The surveyor should undertake records research at the local Recorder's Office or at any other source of land record information. There should be special concentration on deeds or plats that might indicate the establishment of the center of section within the 1856–1883 (or other) time frame. Calls in these deeds for a monument at the center or plats illustrating a center of section or illustrating equal chainage along interior quarter lines would indicate that the center may have been established in accordance

25. Webber, *supra* note 1, at 205; see also the similar comments regarding the 1849–1851 instructions in Brown et al., *supra* note 1, at 318–19 and 374–76.

with instructions contradictory to the 1805 Act. Deeds for parcels that have the center of section as the point of commencement or point of beginning might bear witness to the location, determined by county surveyors, of the center of section.

Any "accepted" center of section created after 1883 must be scrupulously examined since the General Land Office "Rules for the Restoration of Lost and Obliterated Corners," dated March 13, 1883, stated that it was "evident" that the center be located by splitting the quarter sections.[26] Thus, any "accepted" center of section created after this date will clearly be a marked "exception" and by no means a general "rule."

2. The surveyor should perform field reconnaissance and undertake the traversing between all quarter section corners and any monumentation or occupational evidence at or near the center of section; this could bear witness to the possible location of the center.
3. The surveyor should perform office calculations and analysis relating the location of the intersection of the quarter section lines to any single quarter line split or ancient possessory evidence that is in conformity to old deed or plat positions.
4. These calculations and analysis might warrant additional research or extended fieldwork to verify or disprove derived possibilities.
5. The surveyor should contact other surveyors and/or nearby landowners regarding information pertaining to the location of the center of section.
6. After assembling all of the foregoing information and weighing all possibilities, a land surveyor must then decide if there exists an "accepted center of section." Ancient monumentation corroborated by either deeds or plats dated within the twenty-seven-year span is the best proof. Usually this corner location is very close to either the East–West or North–South quarter line split (by distance) and is coincidental with ancient occupation. The final decision as to the location of a center of section is in this manner equitable in a historical perspective.
7. If the surveyor has determined that an "accepted center of section" does indeed exist, a monument record, noting this location, should be prepared and recorded. (Of course, such a procedure should be consistent with state law, if any, relating to monument records).[27] The monument record should not only indicate that the "accepted center" is not at the intersection of the quarter section lines; it should also indicate its location in relation to said quarter line intersection.

---

26. WEBBER, *supra* note 1, at 348 ("From the foregoing synopsis of Congressional legislation [which includes the 1805 Act], it is evident . . . that all subdivisional lines of a section must be straight lines, running from the proper corner in one exterior line to its opposite corresponding corner in the opposite exterior line.").

27. *See, e.g.*, 765 ILL. COMP. STAT. 220/0.01 et seq. (1992), formerly ILL. REV. STAT. 1991, ch. 133, sec. 0.01 et seq.

8. If reference is made in the legal description to the accepted center of section, then it should be inserted into the land description as follows: "Beginning at the accepted center of section, which is _____ feet North (or South) and _____ feet East (or West) of the intersection of the quarter section lines. . . ."
9. A graphic representation of the locational differences should be shown on the prepared plat of survey to further clarify these positions.
10. The acceptable amount of positional tolerances for such a center location has been left to the discretion of the land surveyor because of the unique historical circumstances of each section of land.
11. If the surveyor can find no ancient physical or documentary evidence of the location of an "accepted center of section," then the center of section should be construed as being at the intersection of the quarter section lines.

Note that step 8 above refers to the relationship of the "accepted center of section" and the "intersection of the quarter section lines." Such phraseology was deliberate, as the use of the term "legal center of section," used in describing its relationship with the "accepted center of section," was purposely avoided. One must keep in mind the fact that the center of section is a "record monument" that exists in one place and in one place only.[28] It is thus impossible to have two legal centers of section. Step 8 merely recognizes the fact that if the surveyor who originally divided the section used a procedure that was based on an interpretation of the 1805 Act that did not result in a section divided by intersecting quarter section lines, then the center point produced by said procedure may become the accepted center of section.[29]

In discussing the "center of section" problem with surveyors, one recurring theme emerged—the potential liability of the surveyor for drafting a survey contrary to the 1805 Act. It seems reasonable to state, though, that if the surveyor follows the above eleven steps, any liability is minimized, if not eliminated entirely. By showing all "locational differences" on his plat of survey, has not the surveyor thrust the task of decision making onto the shoulders of the title insurance company? And isn't that where it probably should be? With all facts disclosed, the title company can, frankly, decide where the property lines are. It can take into account such factors as evidence of possession along the disputed area and positional tolerances of center locations. It can then decide if additional underwriting requirements are necessary, such as the exchange of quitclaim deeds or the

28. Madson & Seemann, *supra* note 8, at 77–79, 82, 117.

29. Brown et al., *supra* note 1, at 318–19. Incidentally, the author has in the past seen references on plats of surveys to such lines as "the West line of the Northwest quarter, as monumented." One might consider this heresy, as such a line is also a record monument. However, one might also conclude that the surveyors who prepared such plats were merely, possibly, displaying the same reasoning that is illustrated above in step 8.

execution of a boundary line agreement. Can a surveyor do this? More important, *should* a surveyor do this? Isn't this really the job of the title company?

This article has been an attempt to address the center of section problem from a variety of perspectives. It is hoped that the thoughts contained herein might, in the future, aid the surveyor in reconciling these perspectives when surveying "center of section" lands.

## Acknowledgments

The author wishes to acknowledge the following surveyors who assisted in the writing of this article: Tom Egger, Russ Ory, and John Thornhill, of Donahue and Thornhill, Geneva, Illinois; Jerry R. Broadus of Puyallup, Washington; Shawn VanKampen of Hanna and Associates, DeKalb, Illinois; and especially Mike Filipski of Compass, Ltd., of Naperville, Illinois.

# 18 Record Research
## Paper versus Ground Truth

Wendy Lathrop, PLS, CFM

## I. The Purpose of Record Research

"Record research" has different meanings for the title searcher and the land surveyor. Although each professional may be scouring the courthouse for materials relevant to a particular site, each does it for different purposes and may choose to give varying importance to the documents both will find.

The title searcher is interested in the consistency of the chain of title for a parcel of land. The prime motivation is to determine the marketability of the parcel. Does the title have a doubtful origin, as when a grantor purports to convey a parcel that he or she never owned? Or is the title so burdened with mortgages and other liens that it is essentially without value? Does the wording appear to describe exactly the same property throughout a series of transfers, or do changes in the bearings and distances defining the boundaries give rise to doubt? Is there legal ingress and egress for the site?

The land surveyor is looking at the same documents with an eye to evidence of the parcel's location and configuration. The surveyor searches for clues to the parcel's geometric shape and its relation to adjoining parcels or other physical features such as roads or watercourses. The records are scrutinized for calls to man-made or natural monumentation that may be found in the field, or calls to other documents that may be examined for further descriptive evidence.

The responsibility of both the title searcher and land surveyor is to locate public records relating to the property in question. These records may be found in the county hall of records, but also in the County Surveyor's or Municipal Engineer's office, in local, state, and U.S. federal court offices, local and state tax offices, the Register of Wills or Surrogate's office, and railroad, utility, and other service agencies. Each of these offices maintains records showing the effect of the actions of humanity upon a tract.

In those states where such a position exists, the elected or appointed County Surveyor may "accept for filing maps of surveys that establish monuments and keep a current record of all survey monuments within the county.[1] Thus, a visit to the County Surveyor's office is an important facet of the private surveyor's research. Most of the colonial states do not have such an office, but many of the public land states do.

Other municipal or county offices may hold descriptions of drainage facilities (location of structures, pipe sizes, inverts of inlets and manholes, etc.), drainage and utility easement takings, or plans of lot consolidation or subdivision. Municipal planning agencies may maintain records of minor subdivisions not filed in the county clerk's office.

Of great importance are records of current and historic road width, sometimes referred to as "road returns." In colonial states, these are maps showing the original location and widths, usually in rods, of King's roads, Queen's roads, Great roads, and other highways created either by the sovereignty or local freeholders. Equally significant are ordinances for more recently created or vacated roads, with descriptions of either the created location or the reversionary parcels.

State offices may maintain documents regarding highway acquisitions (which change the boundaries of adjacent properties), highway baseline and monumentation locations (from surveys establishing those roads), and restricted wetland or floodplain areas. Depending upon the scope of services to be provided and the purpose of the survey, the land surveyor may spend as much time in these offices as in the field actually performing the ground survey work.

Land surveyors may search other record sources not usually accessed by title searchers. The original General Land Office (GLO) field notes and surveys are kept in diverse repositories in the various states. Such records may be kept in state library archives, the County Surveyor's office, local or regional Bureau of Land Management offices, or state highway departments. These documents reflect the original markings on the ground delineating the public lands in the rectangular system created by Thomas Jefferson, and are the basis for every subsequent survey in those areas.

Utility company and railroad offices maintain documents showing the location and use of their rights-of-way and easements. Such plans and descriptions are of great use to the land surveyor and the client.

1. Colo. Rev. Stat. 30–20–903(2b).

A few states have statutes establishing registration of land title, generally known as "Torrens systems." In such situations, a designated public official maintains a register of titles. All facts (but not the original documents) relative to the title of a particular property are filed in a registry with this official. Such a registrar may issue certificates of title, which serve as notice of the condition of the title.

## II. Deed Research

Both of our professionals will be looking for deeds as a starting point. The title searcher should look for the source of the title to the property and then check for any claims against each successor in the chain.

Surveyors not only check the cover deed for the site to be surveyed, but also should pull the deeds for all adjoining tracts. One reason for this is that each description expresses an opinion as to the length and direction of a common boundary, and the records must be compared for discrepancies. This may help in the resolution of a deed that does not close, meaning that geometrically following the courses, the description does not end at the same point where it began. Additionally, adjoining deeds may have calls for physical evidence that do not appear in that of the property in question.

In many instances, the surveyor will follow the chain of title back to the creation of a parcel to find additional clues. The most recent deed of record may have been sterilized by a previous attorney or surveyor who wished to simplify or modernize the wording, and thereby left out many important descriptive aspects of earlier versions. Or the current deed may have been recycled for so many years that it has lost its relevance to the real world. Many deeds contain simple typing errors that can cause problems.

The deed of record for the author's home is one such example. On researching the records in 1980 to purchase the property, the current description on file was found to be bounded on the west by the street, running easterly 165 feet more or less along lands of Cadwallader, running southerly 25 feet more or less along lands of Anderson, and running westerly back to the beginning.

This description was repeated through four conveyances, back to 1926. In the interim, all the adjoiners had changed, a boundary line agreement was made along the northerly line, and the southerly line had changed through the combination of a neighbor's garage that had encroached since the 1920s and a prior surveyor erroneously extending the line of the encroaching garage all the way to the rear property line to take an additional two feet from my predecessors in title. Additionally, a common drive easement along the northerly line was called for in the neighbor's deed but never repeated in any description for the author's parcel past the easement's creation.

It should come as no surprise that at settlement, the title company was not ecstatic to see a new description with six courses instead of the old four, nor that some distances did not match the old ones by more than a foot. A call of "through the common party wall" was also added to the document for the first time.

## III. Legal Responsibility—Title Searching Standards

In the title business, the searcher or abstractor gathers relevant public records. These are forwarded to the attorney's or examiner's office, which writes a report for title policy on ownership, subject to easements. In order to close, the attorney must notify the buyer of the report's findings, disclosing any restrictions.

The search must cover all official public records within a stated period of time. Besides deeds conveying the tract, this includes discharge of mortgages, acknowledgment of mortgages, and any restrictions of record. This last factor may include zoning or environmental restrictions affecting the use of the property. The policy may contain coverage exceptions or the statement "if any" regarding restrictions not discovered during the search.

The extent of the title search varies in different localities. In states having a "Marketable Title Act," a person having an unbroken chain of title to real estate under a recorded conveyance for a statutorily prescribed number of years is considered to have marketable title, subject to instruments recorded for less than that prescribed time, or other claims, defects, or encumbrances not barred by the statute of limitations.

Interestingly, in the Uniform Simplification of Land Transfers Act, a time of thirty years is recommended, but the actual times specified by states using this model law vary greatly. For example, Nebraska and South Dakota prescribe twenty-two years for marketable title,[2] while Michigan requires forty years (except twenty years for mineral interests),[3] and Indiana requires fifty.[4]

Additionally, the uninterrupted time for marketable title is not always the same as the statutory limitation of actions for adverse possession. In South Dakota, action is required within twenty years for the recovery of real property or for possession of real property,[5] while in Nebraska, action is required within ten years after the cause of action has accrued.[6]

The point of this discussion is to determine a sufficient number of years to go back into the public records in order to uncover any possible defects or counter claims to title. In states having a marketable title act, the answer is clearly defined by statute. Elsewhere, common practice may be to tie research to the limitation of actions for adverse possession.

---

2. Neb. Rev. Stat. § 76–288; S.D. Codified Laws § 43–30–1.
3. Mich. Comp. Laws § 565.101.
4. Ind. Code § 32–20–31.
5. S.D. Codified Laws 15-3-1.
6. Neb. Rev. Stat. 25–202.

For example, New Jersey does not subscribe to the marketable title act, but sixty-year searches have long been the standard. However, with increased competition, numerous title companies are reducing their searches to thirty years. Both of these practices may be defended by citing statute, as in fact there are two defined periods after which adverse claims ripen. In New Jersey, thirty years of actual possession of real estate, except woodlands and uncultivated tracts, and sixty years for those exceptions, vests complete right and title in the actual possessor.[7] As of this writing (2011), there are legislative proposals to reduce these time frames, but no changes have yet been enacted. Meanwhile, some title companies have reduced the extent of their search, sometimes merely extending prior searches without checking prior histories.

## IV. Legal Responsibility—How the Land Surveyor Uses Deeds

Surveyors are governed by statute or defined practice set by the Board of Registration in each state. Frequently, these various regulations contain guidelines defining the surveyor's responsibility regarding record research. As a typical example, the New Jersey law reads as follows:

> Prior to conducting a survey, the licensed land surveyor shall obtain all pertinent information and documentation in the client's possession relative to the property to be surveyed. Such information may include, but not be limited to, earlier surveys, record deeds, title reports, original tract maps, public records and State, county or municipal maps. When such information provided is not sufficient to meet the owner's needs, the surveyor shall make all reasonable efforts to obtain all information and documentation needed to render an accurate survey.[8]

However, the client cannot be relied upon to have much in the way of relevant reference material. Therefore, the surveyor is obligated to pursue diligent research to recover sufficient and necessary materials. The client may be able to provide a copy of the cover deed for the tract to be surveyed, but the prior illustration should suffice to explain the insufficiency of that single document as the basis of a survey. Adjoining deeds and some history of the property in question also need to be acquired.

In addition to regulations set by the Board of Registration, individual state professional societies may have codes of ethics or standards by which their members abide. These often define minimum research and the responsibility of the surveyor to acquire any relevant documents in the client's possession.

7. N.J. Stat. Ann. § 2A:14–30.
8. N.J. Admin. Code 13:40–5.1(b).

A land surveyor may be required to adhere to the standards jointly promulgated by the American Congress on Surveying and Mapping (ACSM) and the American Land Title Association (ALTA). While usually restricted to surveys of commercial property in past years, ALTA/ACSM standards are increasingly requested for other types of properties as well. Section 4 of the 2011 standards ("Records Research") defines the client's responsibility to the surveyor:

> It is recognized that for the performance of an ALTA/ACSM Land Title Survey, the surveyor will be provided with appropriate data which can be relied upon in the preparation of the survey. The request for an ALTA/ACSM Land Title Survey shall set forth the current record description of the property to be surveyed, or, in the case of an original survey, the current record description of the parent parcel that contains the property to be surveyed. Complete copies of the most recent title commitment, the current record description of the property to be surveyed (or in the case of an original survey, the parent parcel), the current record descriptions of adjoiners, any record easements benefiting the property, the record easements or servitudes and covenants burdening the property (all hereinafter referred to collectively as "Record Documents"), documents of record referred to in the Record Documents, documents necessary to ascertain, if possible, the junior/senior relationship pursuant to Section 6.B below, and any other documents containing desired appropriate information affecting the property being surveyed, and to which the ALTA/ACSM Land Title Survey shall make reference, shall be provided to the surveyor for use in conducting the survey. . . .

Sometimes a client may offer an abstract of title. BLACK'S LAW DICTIONARY defines "abstract of title" as:

> A condensed history of the title to land, consisting of a synopsis or summary of the material or operative portion of all the conveyances, of whatever kind or nature, which in any manner affect said land, including maps, plats, or other aids. . . . Memorandum or concise statement in orderly form of the substance of documents or facts appearing on public records which affect title to real property. . . . [9]

From this, it can be seen that the abstract is the abstractor's opinion of what is most pertinent and that abridgement may delete certain aspects that the surveyor would find significant. Reliance on an abstract means one is depending upon secondary evidence, which is subject to judgmental or accidental omissions and scrivener's errors. Therefore, the surveyor may use the abstract as an aid but should still acquire copies of the original forms of the cited documents.

---

9. BLACK'S LAW DICTIONARY 10 (8th ed. 2004 )

Once obtained, deeds for the property in question and adjoining parcels are usually plotted in a form sometimes referred to as a mosaic or work sheet. This work drawing shows common calls for course direction and length, as well as monumentation. Such a plotting shows consistencies and discrepancies between the various records and also gives the surveyor an idea of where to look on the ground for particular clues to boundary location: pipes, pins, mounds of stones, streams, blazed trees, hedge rows, stone walls, building faces, and so on. A plotting of the chain of deeds may also clarify when, how, and where changes in the boundary occurred.

The same discrepancies in geometric description that frustrate the title searcher may be a clue to the land surveyor of the approximate time of prior surveys and therefore to the age of related monumentation. A knowledge of local surveying customs and the historic use of surveying equipment may explain changes in bearings due to magnetic declination or discrepancies in line lengths due to flat or sloped measurements. Additionally, in 1995, a 12-inch oak from a 1942 deed will be a different size than a tree of similar dimension called for in a deed written in 1883. An iron pipe set in 1902 will likely be rustier and more corroded than one set in 1975. It helps to know what to look for.

Record locations are compared with actual locations on the ground during the fieldwork. Other conflicts may be discovered that do not appear in the written record, however. Record lines of ownership may not match lines of possession. This may indicate overlaps or gaps in title. Evidence of long-time use may give rise to questions of prescriptive and adverse claims or rights.

Upon completion of the fieldwork and computations to determine the relationship of field and record locations of the boundaries of a tract, the surveyor prepares a plan as a graphic report of the facts found. A new description may be written to clarify the present configuration of the parcel and reflect monumentation found or set. Both the new description and the new monumentation act to preserve evidence of the survey. Additionally, the client may be provided with a survey report to explain the procedure and resolution of the survey process.

## V. A Sample Comparison of Deeds for Survey Evidence

The author encountered three very different kinds of descriptions while researching the chain of title for a New Jersey tract and its adjoiners. They are discussed here to illustrate a surveyor's eye to evidence of location, evidence of intent, and the difference between clarity and brevity.

Deed 1 (1815):

> . . . situate in the Township of Amwell, bounded and described as follows: Beginning at a hickory in a line of Nathaniel Larson, corner to Barton

Phillips and thence along said Phillips line (1) North thirty six degrees west, eight chains to a heap of Stones in Samuel Well's Line, corner to John White's Lot, thence along the Same (2) North fifty four degrees east, seven chains and Ninety One Links to a heap of Stones, corner to said Lot in a line of said Rodman's other Lands thence along the Same (3) South thirty five degrees east four chains and thirty six links to a heap of Stones in James Quinby's line thence (4) South twenty eight Degrees and thirty minutes west eight chains and Sixty five links to the point of beginning, containing four acres and eight tenths of an acre of Land.

## VI. Discussion

The description begins with three references—a tie to two adjoiners (whose deeds should now be searched) and a hickory. Unfortunately, the tree size is not given, and this is in a semiwooded area.

Course Number 1 references its location to two adjoining tracts and is terminated by a physical monument. A "heap of stones" may not sound impressive, but in fact, the heaps called for in this description are neatly piled circles, about 6 feet in diameter, of large stones clearly placed by human effort and not the result of geophysical phenomena. The heaps still existed in 1990 when the parcel was resurveyed. Courses 2 and 3 also refer to adjoining tracts that can be found in the public record, and both courses terminate at heaps of stones.

Deed 2 (1929):

. . . lying and being in the Townships of Delaware & Kingwood, in the County of Hunterdon and State of New Jersey, bounded and described as follows, viz:

Lot No. 3—
********** BEGINNING at the southeasterly corner of Lot No. 2; thence (1) along Lot No. 2, North thirty four degrees and ten minutes east crossing the public road and passing on the southerly side of a blazed Ash tree on the easterly side of said road Five Hundred and fifty nine feet more or less to a corner; thence (2) South forty six degrees East one chain and fifty six links to a White Oak tree; thence (3) North eighty eight degrees forty five minutes East four chains and twenty five links to a stone heap on top of hill in line of land now or formerly of Quinby; thence (4) by land formerly of Bernard Rogan South twenty seven degrees and fifteen minutes west one hundred and two feet to a corner in the public road; thence (5) South thirty eight degrees and forty three minutes West ninety one feet too the aforesaid easterly right of way line: thence (6) along said right of way line

> North Fifty five degrees and fifty minutes West, three hundred and fifty nine and fifty four hundredths feet to BEGINNING. Containing five acres more or less.
>
> Being Lot No. 3 in accordance with a partial survey made by Walter E. Roberts C.E. May 1929. . . .

## VII. Discussion

Due to the creation of new municipalities, this description is located in two townships (Delaware and Kingwood) instead of in Amwell, as in the 1815 deed. There is no discrepancy here.

The description starts by labeling our parcel "Lot No.3" and referring to the corner of "Lot No. 2," but we do not know where these lot numbers came from. Do they come from a filed map of subdivision, an unfiled plan of subdivision, or a tax assessor's records? Without this information, we do not know where "Lot No. 2" is in order to begin at its southeasterly corner.

The first course crosses "the public road." What is the name of this road? How wide is it? At what point in the first course do we intersect the road? After crossing the road, we pass "on the southerly side of a blazed Ash tree on the easterly side of said road." This physical call is useful, as evidence of blaze marks often last many years, and we also know what kind of tree we are looking for. The "more or less" distance to the terminus of our course makes life difficult, as there is neither an adjoining tract nor a physical marker to know where to end the line.

The second course changes measurement units from feet to chains and links. This is a clue that we are working with some kind of composite description, one that was not based on a survey of the entire tract. In fact, in the closing paragraph of the deed, we are told that this is indeed a "partial survey." This means that only the new lines to be created were surveyed, not the perimeter of the entire parent tract.

The "partial survey" label also explains why Course Number 1 is a plus or minus distance; it is a new line that goes from the corner of newly created Lot 2 to the perimeter of the parent tract, whatever that distance may be. A new line will have new units of measurement. Since Course Number 2 is in the original chains and links, we know that it is an unchanged course from the original undivided parent tract. The same analysis can be applied to Course Number 3, which is also in chains and links.

Pleasantly, we find physical and record calls at the end of both Courses 2 and 3 (a tree for the former and a heap of stones plus an adjoiner's tract for the latter). Additionally, we find a record link between this tract and the one in the previously discussed deed so that we can begin to plot our deed mosaic: "a stone heap on top of hill in line of land now or formerly of Quinby." This is also more definitive in terms of terrain, and we now know we need to look on a "hill" for the stones. In fact, this "hill" is a ridge rising about 70 feet from the roadway, and this particular stone heap was found in the field almost immediately.

Course No. 4 reverts to feet, so we know we are once again running along a newly created line. Somehow, we end at "a corner in the public road," which gives rise to more questions. Which public road are we in? Is "in the public road," a reference to the center line of the right-of-way or to the traveled way? Or are we in fact just somewhere in the road, not necessarily at the center of anything? (In this case, the line was in fact a line run diagonally across the road.) Statutes or local custom may assist in the interpretation of deed verbiage.

The fifth course does run to the right-of-way line of the road, and although we still do not know the width of that right-of-way, by the phrasing, we now know definitively that the blazed Ash in the first course is on the right-of-way line and not just the side of the traveled way. Our fifth course is another new line, in feet and hundredths of a foot, returning to the tract's beginning point.

Deed 3 (1978):

ALL THAT CERTAIN described land located in; Township of Kingwood, County of Hunterdon, and State of New Jersey more particularly described as follows:

BEGINNING at a point for a corner Northerly of a road known locally as Stompf Tavern Road, said corner being the Northeast corner to lands of Carl Woronowicz; thence

(1) North 57 degrees 16 minutes 45 seconds East a distance of 220.00 feet to a point for a corner; thence

(2) South 28 degrees 25 minutes 20 seconds East a distance of 447.35 feet passing over Stompf Tavern Road and running along lands remaining to Robert Chabot to a point for a corner; thence

(3) South 57 degrees 16 minutes 45 seconds West a distance of 220.00 feet along Chabot to a point for a corner in a line of lands of the aforementioned Carl Woronowicz; thence

(4) North 28 degrees 25 minutes 20 seconds West a distance of 447.35 feet along Woronowicz and passing over Stompf Tavern Road to the point and place of beginning.

CONTAINING 2.259 acres of land. . . .

## VIII. Discussion

This is the only one of the three deeds that closes mathematically, but it is also the least helpful in terms of determining its location on the ground. Most obvious is that each course merely runs "to a point," with little in the way of physical clues. Looking more closely, other questions arise.

We begin at a point "northerly of a road"; how far northerly? Are we northerly of the physical traveled way or of the recorded right-of-way? What is the right-of-way width at the time of this recording? This road is "known locally as Stompf Tavern Road"; does it have other more generally known names? How local is local? We are at the "Northeast corner to lands of Carl Woronowicz"; Mr. Woronowicz is quite the land speculator; which of his many tracts are we referred to?

Course Number 1 gives no clue as to ties to adjoiners or relation to roads, monumentation, etc. Course Number 2 passes over Stompf Tavern Road, but we do not know at what point in relation to the beginning or terminus of this course such an intersection occurs. Fortunately, we have a call to lands remaining to Robert Chabot, but we need to find his deed to find out where his property lies as well. Course Number 4 terminates in a line of lands of our land speculator, so we are not certain which of his lands we have run into; we have in fact been fortunate to encounter the same parcel at which we began. Course Number 4 again does not specify where the course crosses Stompf Tavern Road.

As a last comment on Deed 3, note that all numbers are written in numeric form only, without alpha spelling. The possibility for transposition or outright blunder is more common with numerals. The author prefers holding out written words over numerals when a discrepancy exists.

## IX. Comparative Value of These Deeds to the Surveyor

Although the geometry of the 1815 deed is the worst, the descriptive aspects of this document made it the most valuable of the three deeds. The clear calls for physical monumentation made those features readily identifiable in the field. Calls for adjoining owners also simplified further record research. The geometry and mathematical aspects of the 1815 deed are easy to correct and update, but other descriptive features cannot be replaced if lost over the years.

## X. Conclusion

The concerns of the title searcher are related to the marketability of the property: proper parties to convey, adverse claims, clear title, consistency of description, and encumbrances. Insurability seems to be the great motivator.

In contrast, the surveyor is looking for written evidence of property location: geometric description of lot bounds (bearings and distances), calls for physical evidence, calls for adjoiners (which may require pulling more deeds), and calls for maps (which become part of the description if their location is clearly referenced). Additionally, the surveyor is licensed to protect the public and is required to provide service to benefit the lay person. Clarification and improvement of the record is one aspect of this responsibility.

The result of record and field survey work may be a new description intended to clarify the current status of the parcel. Repeating an old description merely to maintain consistency in the record creates problems for future researchers, whether of title or survey persuasion. Such complications require additional time to resolve and therefore incur further expense to the client.

## *Additional References*

Uniform Simplification of Land Transfers Act, 14. U.L.A. 209; USOLTA § 3–301 et seq. (1980) ("Marketable Record Title," as amended to 1990).

Jay M. Zitter, J.D., Annotation. Construction and effect of "marketable record title" statutes. 31 A.L.R. 4th (1984).

W.W. Allen, Annotation. Construction of "marketable title," "ancient claims extinguishment," and like statutes, terminating or limiting enforcement of rights, interests, and claims respecting real property, based on old records, instruments, or events. 71 A.L.R.2d 846 (1960).

Thomas G. Fischer, J.D., Annotation. Specificity of description of premises as affecting enforceability to contract to convey real property-modern cases. 73 A.L.R. 4th 135 (1989).

# Locating Utility Easements in New York State

## A Detailed Analysis[1]

19

Gerald J. Greenan, Esq.

Lawyers (and surveyors) are frequently asked to locate utility easements on property being sold or financed by their clients. In the context of a sale, responsibility for locating and evaluating an easement is primarily that of the purchasers' attorney. When the sellers' attorney is asked to locate an easement, it is usually because the public record does not permit the examiner to determine whether improvements on the premises can be disturbed by the owner of the easement.

## I. Definitions and Distinctions

BLACK'S LAW DICTIONARY (abridged 6th ed. 1994) defines an "easement" as "[a] right of use over the property of another" and "right of way" as a "[t]erm used to describe a right belonging to a party to pass over land of another . . ." (In this discussion I use the terms interchangeably.) Please note that an easement is either the name of an instrument granting rights in the lands of another or the rights themselves.

An easement is not a pipe line, a pole line, or an underground conduit. Pipe lines, pole lines, and underground conduits are physical facilities installed in the property of others by virtue of rights

1. An earlier version of this article appeared in the 33(1) N.Y. REAL PROP, L.J. 28 (2005).

granted by an easement. The distinction is important. Usually, when someone asks a lawyer or surveyor to locate an easement, what they really want to know is where the pipes are buried.

## II. Don't Ask

It is not proper to ask a lawyer or surveyor whether a pipe is buried someplace on a piece of property. He or she does not usually know, and if he or she did happen to know, it would not address the question of whether a pipe was there as a matter of right.

In western New York, to determine the physical location of underground pipes, or whether it is safe to dig, the wise contractor or homeowner does not call his lawyer or surveyor. He calls "Before You Dig" (800–962–7962). One call results in all local utilities marking their underground facilities.

It is proper to ask a lawyer (although not a surveyor) whether an easement permits a pipe to be buried in the future through an about-to-be constructed dwelling or whether a dwelling built over a pipe line can remain as a matter of right.

## III. Undefined Easements

Frequently, an examination of the easement instrument reveals that neither party knew exactly where the final poles or pipes were to be installed, and neither thought it critical at the time the instrument was executed.

We often see easements given by Farmer Brown that are no more definite than the following: "Bounded north by Bebak, east by Smith, west by the Town line, and south by Clinton Street." Does the language tell which of Farmer Brown's three parcels on Clinton Street is to be subject to an easement? Does it define the boundaries of the easements, or does it simply identify the parcel?

It is "black letter" New York law that the language of an agreement must be construed most strongly against the party who prepared it or supplied a form for the agreement.[2] But reference to general principles of law is not enough to convince most title attorneys, who tend to be (and are paid to be) ultraconservative. The location of an easement described as above is, arguendo, an easement over all of the property in question—what is frequently called a "blanket" easement.

2. *See* 22 N.Y. Jur.2d, Contracts § 260.

## IV. Blanket Easements

I wince whenever I see a reference to a "blanket" easement. The term is misleading. "Blanket" implies that it covers or affects all of the property. "Undefined" is the correct adjective to describe these easements.

Does having a "blanket" easement imply that a utility company has the right to install pipes or poles any place on the property at will or at whim at any time between the date of the instrument and the date of the premises being released from the easement? No!

Facilities installed pursuant to an indefinite easement are to be installed where the owner of the property wants them to be installed, and not randomly over the entire property described in the easement. This is well-settled law, although not well-known law.

## V. The *Onthank* Doctrine

The seminal case on easements is *Onthank v. Lake Shore & Michigan Southern Railroad Co.*, 71 N.Y. 194 (1877).

A discussion of the topic is found at 49 N.Y. JUR.2D (rev.), EASEMENTS §§ 94–99. I quote omitting citations and inserting captions, parenthetical material, and paragraph spacing for clarity:

> § 94. Where the grant of an easement of way does not definitely locate it, a reasonable and convenient way for all parties is implied, in view of all the circumstances.

### *A. First Right to Locate*

> If an easement of way is created but not located by a grant, the owner . . . has, in the first instance, the right to designate the location, provided he or she exercises that right in a reasonable manner, having regard to the suitability and convenience of the way so located to the rights and interests of the owner of the dominant tenement (e.g., the utility company).

### *B. Second Right to Locate*

> If the owner . . . fails or refuses to locate the way, the grantee acquires the right to make his own selection of a location, having due regard to the interests, rights and convenience of the other party (the owner).
>
> An easement granted in general terms, without giving definite location and description, so that the part of the land over which the right is to be exercised cannot be definitely ascertained, does not give the grantee a

right to use the servient estate without limitation as to the place or mode in which the easement is to be enjoyed.

### *C. Grantee Cannot Relocate*

§ 98. Where a granted easement has been exercised in a fixed and definite course, with the full acquiescence and consent of both parties, it cannot be changed at the pleasure of the grantee.

### *D. Refining Location*

§ 99. Ordinarily, a grant or reservation of a right-of-way "over" a particular area, strip, or parcel of ground is not to be construed as providing for a way as broad as the ground referred to. The grant of a right-of-way "through and over" a space 20-feet wide is the grant of a convenient way within those limits. [The authority for this statement is Benjamin Cardozo's Court of Appeals in *Dalton v. Levy,* 258 N.Y. 161 (1932). In the *Dalton* case, the Court of Appeals permitted a garage built 9 feet into a 20-foot right-of-way. 258 N.Y. at 167.]

## VI. Grantor May Relocate Access

Although the *Onthank* decision of the Court of Appeals was handed down in 1877, it is still the law in New York, as evidenced by the court's recent decision in *Lewis v. Young.*[3]

In *Lewis v. Young,* the court of appeals, citing and referring specifically to *Onthank,* distinguished the one-time right to locate a pipe from an undefined right of ingress and egress over the land of another. The court reached the conclusion that an undefined right of ingress and egress is a right of passage and not a right in the physical passageway itself.[4]

The right to relocate is given the servient estate (landowner) without consent of the dominant estate, conditioned upon the relocation being without expense to the dominant estate.[5]

*Lewis v. Young* is an extremely useful case. Even the most well-defined easements customarily have appurtenant undefined rights of ingress and egress for the purpose of repair and replacement.

3. 705 N.E.2d 649 (N.Y. 1998).

4. 705 N.E.2d at 652 (citing Bakeman v. Talbot, 31 N.Y. 366, 371 (1865)).

5. In a typical modern subdivision, the area reserved for utility easements is relatively definite. It is either shown on a map made a part of the easement agreement or on the filed "Map Cover," or it is defined in the instrument, e.g., "within 10 feet of the south line of subdivision lots 20 through 26. . . ."

If private attorneys are generally unaware of the *Onthank* Doctrine, the same cannot be said of the attorneys for public utility companies. In the western New York area, the two local electric utilities have historically taken opposite approaches to the *Onthank* Doctrine, and the approaches are familiar to experienced local title examiners.

## VII. New York State Electric & Gas Corporation

The New York State Electric & Gas Corporation obtains and records a new easement for every new service line outside a modern subdivision. An examination of a title will frequently disclose three or four practically identical indefinite easements to New York State Electric & Gas Corporation granted prior to inclusion of the property in a modern subdivision. New York State Electric & Gas Corporation knows that, under the *Onthank* Doctrine, once the original line has been installed, the same instrument cannot be used to install new lines in a different physical location. It therefore obtains a new easement agreement.

## VIII. Niagara Mohawk Power Corporation

Niagara Mohawk Power Corporation, with the same knowledge, frequently did not bother to incur the expense of recording its easements. They cannot be used for future expansion, and once a pole line is installed, recording is not necessary.[6]

Easement rights to maintain the original pole line are protected by the notice afforded by the physical presence of the poles and wires on the property: "the presence of overhead wires strung from poles and crossing land in clear view was sufficiently open, visible and obvious upon even a cursory examination of the property to constitute constructive notice, to a reasonably careful and prudent buyer, of the existence of some right or claim of right to possession by those maintaining such power transmission line." (citations omitted)[7]

## IX. Resolving Encroachments

If a reference to the *Onthank* Doctrine and its progeny does not resolve the issue of locating an easement and notwithstanding the *Onthank* Doctrine, the printed form of contract most widely used in upstate New York provides that title will be accepted subject to

6. When Niagara Mohawk was acquired by The Grid, it became company policy to record future easements.

7. WARREN'S WEED NEW YORK REAL PROPERTY § 20.01, *Notice of Easement* (4th ed. 1990).

> easements and rights-of-way of record for water lines, sanitary sewer lines, drainage, gas pipe line, electrical lines and telephone lines, provided they are or may be used to service the property and provided the present improvements . . . are or will not be *on the easements or rights of way* . . ." (emphasis added)

I counter with Real Property Actions and Proceedings Law section 2001.

## *A. § Real Property Actions and Proceedings Law Section 2001 and the Oneida Doctrine*

Most attorneys dealing in real estate are familiar with the use of Real Property Actions and Proceedings Law (RPAPL) section 2001 to "cure" violations of building restrictions by establishing that a violating structure has been in place for two years or more.

Many do not realize that the court of appeals has construed the statute as applying to all negative easements, not only those contained in building restrictions.[8] A negative easement is one that gives the dominant estate the right to tell the servient estate not to do something; for example, "don't build under our wires."

The language of the Niagara Mohawk easement in the *Oneida* case does not appear in the opinion of the court of appeals, but it does appear in the lower court's opinion.[9] The granting clause is as follows:

> [T]he grantor has granted and released and does hereby grant and release . . . a right of way and easement to build, rebuild, relocate, operate, repair, maintain and, at their pleasure remove a single line of so many poles . . . as either of the Companies may now or shall from time to time deem necessary, and to transmit and distribute electricity and render telephone and telegraph service upon, over, across, through, under and beyond the land, including land within the adjoining highways, which the grantor owns or in which the grantor has an interest.
>
> With the right to trim, cut and *remove any trees, limbs, brush or other obstructions* on either side of said lines which either of the Companies may deem likely to interfere with the operation thereof.[10]

The objection of Niagara Mohawk was that mobile homes were located directly beneath its 9.6 kilovolt wires, a position deemed by Niagara Mohawk to be threatening to the lives of those who lived in the mobile homes. Even though the granting clause was indefinite as to location, the mobile homes were clearly encroaching.

---

8. Oneida Cnty. Mobile Home Sales, Inc. v. Niagara Mohawk Power Corp., 393 N.E.2d 1024 (N.Y. 1979).

9. Oneida Cnty. Mobile Home Sales, Inc. v. Niagara Mohawk Power Corp., 407 N.Y.S.2nd 89 (N.Y. App. Div. 1978).

10. *Id.* at 89 (emphasis added).

The mobile homes had been in position for more than two years (RPAPL§ 2001), but less than the ten years (N.Y. C.P.L.R. § 212) required to establish prescriptive or adverse use and less than the six years in which an action could be brought to remove an obstruction that interferes with an affirmative easement (N.Y. C.P.L.R. § 213(1)).

The court of appeals found that RPAPL section 2001 applied and imposed on Niagara Mohawk the cost of moving the wires or mobile homes.

The lower courts have had great difficulty with the negative-affirmative easement distinction found in *Oneida.* They sometimes forget or ignore the fact that the court in *Oneida* ordered the pole line moved at Niagara Mohawk's expense.

See, for example, *Rahabi v Morrison,*[11] in which the Appellate Division, Second Department, permitted a suit to continue beyond the two-year period when the suit was for the removal of a chain link fence that interfered with easement access to plaintiff's property.

The distinction between affirmative and negative easements becomes moot if the "encroachment" on the easement is more than six years old.

## X. Not a Statute of Limitations—A Release

It is interesting and significant to note that section 2001 is not a statute of limitations. Subdivision 5 provides: "If an action governed by this section is not commenced within the time herein provided it shall be *conclusively presumed* that the right of action for the relief for which that action might have been brought has been *released.*" (emphasis added)

The Law Revision Commission, in recommending the passage of the statute, commented:

> Short Statutes of Limitation have been enacted in some jurisdictions limiting actions for enforcement of any restriction on the use of land. . . . A bar framed as a Statute of Limitations, however, is inadequate to solve the problem, since the period may in a particular case, be tolled by disability of a person entitled to enforce the restriction or by nonresidents or absence from the state of the party against whom relief is sought. Uncertainties may also be injected if it is asserted that the bar of the statute has been waived. . . .
>
> Legislation creating a conclusive presumption of release of the right of action, if action is not brought within the time limited, would provide assurance free of these uncertainties.[12]

11. 440 N.Y.S.2d 941 (N.Y. App. Div. 1981).
12. (1963) N.Y. Sess. Laws 1814. (McKinney),1816.

## XI. Improvements More than Two Years Old

Whether the area subject to an easement is well defined or undefined, improvements in place for two years may remain undisturbed pursuant to the *Oneida v. Niagara Mohawk* Doctrine, and the area of the improvements is subject to the conclusive presumption of release of RPAPL section 2001.

Since the easement rights are conclusively presumed to have been released, it is inappropriate to demand a further release, and the "present improvements" cannot be deemed "on" or encroaching on the easement under the terms of a "standard contract." The location of improvements in place for more than two years does not affect contract marketability.

## XII. Improvements Under Two Years Old Aboveground

If the area subject to an easement is defined by the instrument creating it—for example "within the south 25 feet of premises"—then a request to "locate the easement" is not a reasonable request. (The examiner should look at the map of survey and draw his or her own conclusion as to how property is affected.)

If the area subject to an easement is truly indefinite (a "blanket easement"), then a request to "locate the easement" is not a reasonable request when made to an attorney or surveyor. Neither is licensed to change the language of a recorded instrument from indefinite to definite. The easement area is governed by the *Onthank* Doctrine and, for aboveground facilities, is "located" by reviewing a current survey.

## XIII. Belowground

For underground facilities, the survey will give only "clues"—for example, storm and sanitary manholes, cathodic protection for gas transmission lines, backyard drain inlets, telephone pedestals, and electric junction boxes.

Since neither the attorney nor the surveyor is equipped with X-ray vision, each public utility has a "Call Before You Dig" telephone number. A call to that number will result in each utility marking the area where their records and equipment indicate underground pipes or wires are contained.

## XIV. Future Exercise of Easement Rights

But what about the future exercise of easement rights?

If the area is well defined by the instrument, its terms may prevail if it conflicts with a structure completed for less than two years.[13] If the area is undefined by the instrument and its location has not been established by the previous installation of facilities, then the *Onthank* Doctrine prevails. Facilities installed pursuant to an indefinite easement are to be installed where the owner of the property wants them to be installed, and not randomly over the entire property identified in the easement.

## XV. Never Ask?

Does all of the above mean that it is never appropriate to ask a sellers' attorney or surveyor to locate an easement?

No. It means that it is never appropriate to ask a sellers' attorney or surveyor to locate an easement without first examining the instrument and determining whether it is definite or indefinite. If the easement is definite and its location cannot be plotted without information outside the historical title (for example, widening a right-of-way by the Department of Transportation using physical monuments), then asking a surveyor to locate it using his professional access to records outside the County Clerk's Office is appropriate.

If the easement is indefinite, the next question is whether it became defined by the installation of facilities shown (or not shown) on the map of survey.

If it is aboveground, look up. If it is belowground (at least in western New York), call 800–962–7962.

If presented with an indefinite easement for a pole line and with an accurate survey showing no pole line, the conclusion under *Onthank* is that the pole line, if it exists, is located outside the parcel surveyed. It is not appropriate to ask a sellers' attorney or surveyor to locate pole lines outside the parcel surveyed. If the pole line does not yet exist, then the conclusion under *Onthank* is that it is yet to be located. Surveyors and attorneys are not licensed to predict where lines will be placed in the future. This does not create a problem, because under *Onthank* and its progeny, the owner has the first right to designate a convenient location.

## XVI. Conclusion

Onc who professes ignorance of the location of an easement may be confessing that one did not cxaminc the instrument, or may be confessing that one does not understand the law of easements.

Both are easily forgiven so long as they do not transfer the burden of the ignorance to another attorney.

13. See, however, the "convenient way" discussion under the heading "Refining Location" above.

# 20 Uncertainties in Boundary Locations

Gary R. Kent, PS

In the process of conducting a boundary survey, a variety of conflicts between deed descriptions, plats, surveys, and even evidence on the ground may occur. An exploration into the causes of these potential discords is warranted since they often result in apparent differences of opinion between surveyors. Additionally, many conflicts between legal descriptions have their genesis in the surveys that were performed as a part of the creation of those descriptions, thus a look at the sources of uncertainty in boundary surveying is appropriate.

There are four potential sources of uncertainty in boundary locations. These result from (1) uncertain, indeterminate, or conflicting reference monuments; (2) inconsistent, erroneous, or conflicting legal descriptions and survey plats or maps; (3) lines of occupation or possession that differ from the lines of title described in the deed; and (4) the inherent imprecision in any measurement.

Sections 3.E.ii. and 3.E.iii. of the new 2011 Minimum Standard Detail Requirements for ALTA/ACSM Land Title Surveys state the following:

> 3.E.ii.—Any boundary lines and corners established or retraced may have uncertainties in location resulting from (1) the availability, condition, history and integrity of reference or controlling monuments, (2) ambiguities in the record descriptions or plats of the surveyed property or its adjoiners, (3) occupation or possession lines as they may differ from the written title lines, and (4) Relative Positional Precision. Of these four sources of uncertainty, only Relative Positional Precision is controllable, although due to the inherent

errors in any measurement, it cannot be eliminated. The magnitude of the first three uncertainties can be projected based on evidence; Relative Positional Precision is estimated using statistical means (see Section 3.E.i. above and Section 3.E.v. below).

3.E.iii.—The first three of these sources of uncertainty must be weighed as part of the evidence in the determination of where, in the surveyor's opinion, the boundary lines and corners of the surveyed property should be located (see Section 3.D. above). Relative Positional Precision is a measure of how precisely the surveyor is able to monument and report those positions; it is not a substitute for the application of proper boundary law principles. A boundary corner or line may have a small Relative Positional Precision because the survey measurements were precise, yet still be in the wrong position (i.e. inaccurate) if it was established or retraced using faulty or improper application of boundary law principles.

## I. Reference/Controlling Monumentation

Any boundary survey must begin at, or be based on, at least two known "control" or "reference" points. Boundaries are not established out of thin air; they must relate in some way to other known locations, such as street rights-of-way, section lines, or grant corners.

Reference monuments can take many forms since there can be many points and lines on which the boundary is dependent referenced in any given description. The location of each of these dependent points or lines must be recovered, established, or otherwise determined in order to retrace the boundary properly and with integrity.

Frequently, however, the referenced point or line is not well documented. For example, a description may call for a line to run parallel with a railroad right-of-way line. If that particular railroad was abandoned thirty years ago, determining exactly where the right-of-way line was may be quite problematic. In fact, it is highly unlikely that two surveyors working independently would come up with exactly the same location. This is because of imperfect (conflicting or missing) evidence on which the location is to be based.

Likewise, a description may commence at the intersection of the centerlines of two old county or township roads. The surveyor may find few, if any, records related to that location. Upon field investigation, perhaps no marker was found, in which case the location must be reestablished—often based on conflicting, old, and/or incomplete records and evidence. The resulting location will not have as high of a level of integrity or "confidence" as if the original marker had been found.

Different surveyors may interpret the records and evidence differently, thereby coming up with their own location of the same point. Obviously, surveys utilizing different beginning points will result in boundary corners and lines that are potentially in conflict.

## II. Record Documents

Another source of boundary conflicts lies in the records themselves. Many boundary lines, where one would expect neighboring properties to share "common" lines, in fact have gaps and overlaps based on their respective legal descriptions. The exact source of these variances can sometimes be found and sometimes not. They may be the result of unqualified persons writing descriptions or of descriptions that were written without the benefit of a land survey.

Sometimes a problem is simply the result of a mistake that occurred in the transcription of a deed and that has been perpetuated through the years. Sometimes these problems are the result of descriptions based on conflicting surveys (e.g., as discussed above under "Reference/Controlling Monumentation." Conflicts in the records may even be the result simply of poor survey work in the first place.

In addition to gaps and overlaps, many old plats do not provide enough dimensional information for a surveyor to mathematically resolve the various lot corners and lines. Many old plats show irregularly shaped lots yet only give one or two dimensions, leaving it to the retracing surveyor to try to sort it out. Again, two surveyors coming to the same definitive solution in such instances is nearly impossible.

In any case, in a boundary survey, it is important to recognize that conflicts and irregularities can occur in the records themselves, and this is not unusual or unexpected.

## III. Lines of Possession or Occupation

Oftentimes in the process of performing a boundary survey, lines of occupation or possession such as fence lines or tree rows will be found near, but not exactly on, the actual boundary line. Sometimes, these lines will be significantly different from the line as described in the deed.

This is important to recognize since great weight is often placed on lines of occupation—sometimes justifiably, sometimes not. Often, fence and tree lines, for example, are reliable evidence of where a boundary may lie. But sometimes, they differ significantly from the line described in the deed, in which case they may simply be a fence or tree line.

These differences may also be evidence of the movement of lines by unwritten means such as adverse possession or acquiescence. But finding the solution to such issues requires a deep understanding of boundary law. Sometimes, conflicts can be avoided or mitigated by encouraging the owners to establish an agreement line, but sometimes a quiet title action may be required.

In any event, the mere existence of a line of possession or occupation does not mean that it defines the boundary line. Generally, for possession to be considered evidence of original survey lines:

- There must have been a controlling survey that, if located, would control the lines between the adjoiners.
- Lines of possession are along the lines surveyed or are presumed to have been surveyed by the original surveyor.
- A series of possessions in agreement with one another will tend to substantiate one another.
- The possession must be of a former generation (ancient) or testimony can be taken as to its origin.
- Possession has the reputation of being on the correct survey lines.[1]

## IV. Measurement Uncertainty

There is no such thing as a perfect measurement. No matter what type of measuring device is used, the "true" length of anything is unknown—at least within the measurement tolerance of the device being used. Surveying is no different.

Surveyors have many highly accurate measuring devices, such as electronic total stations, EDM, lasers, and GPS. But none of these are perfect, and all result in some level of uncertainty in the measurement. When current technologies are applied in land surveys to retrace boundaries that were originally surveyed with a compass and chain in the mid-1800s, these measurement issues can become significant.

In addition, surveyors are charged with making sure that their staff are appropriately trained. They know that they must compensate for systematic errors in their measurements, including those associated with instrument calibration, or their results will not be precise. They know that they must use appropriate error propagation and measurement design theory, which, in part, involves selecting the proper instruments, geometric layouts, and field and computational procedures so that the random errors that are inherent in any measurement are controlled to the extent possible and practicable.

The criteria by which surveyors are to evaluate the integrity of their measurements is called Relative Positional Precision (RPP) and it is defined and explained in sections 3.E.iv. and 3.E.v. of the 2011 ALTA/ACSM Standards as

> the length of the semi-major axis, expressed in feet or meters, of the error ellipse representing the uncertainty due to random errors in measurements in the location of the monument, or witness, marking any corner of the surveyed property relative to the monument, or witness, marking any

1. Brown, Robillard, & Wilson, Evidence and Procedures for Boundary Location 111 (5th ed. 2006 ).

other corner of the surveyed property at the 95 percent confidence level (two standard deviations).

RPP uses the rules of error propagation to analyze measurements and develop what should be the best estimate of the true locations of the points determined on the survey. The maximum allowable "imperfection" in measurements—the length of the semi-major axis—on a 2011 Land Title Survey is 2 cm (0.07 feet) plus 50 ppm. On a survey of a typical 1-acre tract, this would amount to about 1 inch.

## V. Summary

Interestingly, of these four sources of uncertainty in boundary location—reference monuments, record documents, lines of possession/occupation, and measurement—only the measurement uncertainty is controllable by a surveyor; and even so, it cannot be eliminated, only minimized.

The first three sources of uncertainty are part of the evidence that the surveyors use in forming opinions about where boundary lines or corners should be placed. The measurement uncertainty is a function of how precisely those locations can actually be marked on the ground.

One or more of these four sources of uncertainty very frequently results in apparent disagreements between surveyors. Such disagreements can often be resolved by having the two surveyors sit down and discuss their surveys and the evidence they relied upon. However, sometimes differences of opinion as to the weight that should be given certain pieces of evidence, the interpretation of a particularly vexing description or old plat, or the impact of an old line of occupation as evidence of the original line simply cannot be readily resolved and the two surveyors will remain in disagreement.

Armed with at least a basic understanding of these potential sources of such disagreements and in conjunction with the surveyors involved, attorneys will be in a position to guide the parties, when possible, to a mutually acceptable solution short of litigation.

21

# *Aerial Photography in the Land Development Process*
## *Understanding the Basics of the Land Surveyor's Role*

Landon Blake

## I. Introduction

This article is about the use of aerial photography in the land development and land management process. This discussion will include:

1. Traditional uses of aerial photography in land development and land management.
2. How aerial photographs are acquired.
3. The limitations of aerial photography.
4. The surveyor's role in the acquisition of aerial photographs.

For the purposes of this article, we will define aerial photography as the process of acquiring photographs of the Earth's surface using aircraft and cameras specially designed for that purpose. The aerial photographs are supplemented by precise measurements of features visible in the photos and the location of the camera during the flight. These supplemental measurements allow the photographs to be used as a basis for topographic mapping. The process of producing measurements and mapping from aerial photographs will be defined as aerial photogrammetry.

## II. Uses of Aerial Photography in Land Development and Land Management

We will examine seven uses of aerial photography.

1. Aerial photos can be used by land surveyors to locate topographic features that control the location of a parcel boundary. Consider the following examples: (a) a parcel boundary located along the center of a stream, (b) a parcel boundary located along a ridge line, and (c) a parcel boundary fixed 10 feet from the toe of a levee. In many cases, it is quicker to extract the location of these features from topographic mapping prepared from aerial photography than it is from data collected from ground surveys.
2. Aerial photography can be used for survey planning. Available aerial photography may not be of good enough quality or sufficient resolution to produce a project's required mapping products. However, it can often be used to plan the execution of more accurate surveys. The aerial photography used for survey planning can assist in determining how the project site will be accessed, the limits of vegetation or structures that will impact survey execution, the type of terrain, and the number and complexity of topographic features that may need to be included in the survey.
3. Land surveyors will approximate water boundaries and other complex boundaries using a meander line. The meander line can then be used to calculate areas and represent the boundary graphically. Topographic mapping prepared from aerial photography can be used to create these meander lines without the need for a ground survey.
4. Aerial photography can be used by land surveyors to identify physical features like fences and buildings that encroach on parcel boundaries. The type of encroachment and its approximate physical extent can also be determined from the aerial photography.
5. Evidence of property use by "foreign parties" can be identified on aerial photography. Examples of this type of use are roads over the subject parcel used to access neighboring parcels and the planting of crops across the parcel boundaries.
6. Land development projects can be impacted by environmentally sensitive sites (for example, riparian habitat along a creek or vernal pools), sites of contamination, and historic resources. Aerial photography can be used to make a preliminary identification of these sites.
7. An understanding of topography and land use is needed for most studies of project feasibility and the project planning that follows it.

Aerial photography is often used to obtain this data for project planning purposes. For example, aerial photography could be used to determine the preliminary alignment of a new road. The aerial photography would allow road designers to reduce impacts to homes and other structures.

## III. How Aerial Photography Is Acquired

Aerial photographs are acquired in a process involving six main steps. A seventh step produces topographic mapping from photographs.

In the first step, land surveyors establish survey control that will be used to locate the aerial photograph in a coordinate system. This typically includes determining accurate coordinates on aerial targets that will be visible in the photographs. It may also include determining accurate coordinates on other topographic features visible in the photos. In certain applications, land surveyors assist with establishment of survey control that will be used to precisely locate the camera during the flight.

The second step in the process requires that a plane carrying one or more cameras be put in the air over the project site. These cameras are specially designed for aerial mapping photography. The plane may also carry GPS receivers to help determine the position of the camera when each aerial photograph is taken. A great deal of planning is needed before the plane can leave the runway. The direction of flight, spacing between flight lines and exposures, as well as the altitude of the plane over each point on the flight line must be calculated ahead of the actual flight. This ensures the photos have the correct orientation, scale, and overlap with adjacent photos.

In the third step of the process, the aerial photos are typically obtained as, or converted to, digital photos. Negatives from film cameras are scanned on special high-resolution scanners. Digital photos can be obtained directly from more modern digital aerial cameras. The photographs that come from the camera cannot be used directly as a map background. The aerial photos may have unwanted tilt and relief displacement that prevent their use as a map.

Because a camera image plane is not perfectly level with respect to the ground surface below, all aerial photographs have some degree of tilt. As a result, measurements between two points made directly from the photograph "as taken" may be longer than the actual horizontal distance between the points. In a process known as simple rectification, this tilt is removed from the aerial photographs. This is the fourth step.

In the fifth step, topographic mapping is produced from the aerial photography. The topographic mapping is typically made up of two parts. The first part is planimentric mapping, or creating a two-dimensional map representing the footprint of features like buildings, roads, parking lots, and vegetation. The second part is creating a three-dimensional mapping, which typically includes a three-dimensional surface model or digital terrain model of the project site. Contour lines, spot elevations, and drainage characteristics can then be derived from these lines.

All objects in aerial photos, such as buildings or trees that are above or below the elevation at the center of the photo will show relief displacement. This is the name of the visible lean of objects in an aerial photo and the reason you can see the sides, and not just the top, of tall objects in the photo. Relief displacement even affects the ground surface or terrain in an aerial photo. The more varied and rugged the terrain, the greater this relief displacement will be. Some of the effects of relief displacement can be removed through the production of orthophotos from aerial photos. Orthophotos can be used more like a map because the effects of relief displacement are reduced. Orthophotos may be produced as part of the sixth step. (An existing surface model is needed to produce orthophotos. If a surface model is not produced as part of the fifth step, another source of elevation information is required to produce the orthophotos.)

## IV. The Differences between Satellite Photography and Aerial Photography

This article merits a brief discussion on the significant differences between aerial and satellite photography. The main difference between the two types of photography is the camera platform, or the vehicle used to carry the camera when photos are taken. Most aerial photography is done using an airplane (although helicopters, balloons, or even kites can be used). Satellite photography is acquired from a camera carried in a space vehicle orbiting the Earth.

The difference in the camera platform results in important differences in the resulting photography. Because the camera platform in satellite photography is much farther from the surface of the Earth, the resulting photography is typically of a smaller scale and lower resolution than aerial photography. This means each satellite photograph typically shows more area in less detail, and each object in fewer pixels than an aerial photograph of the same area. Because a satellite is much higher above the Earth's surface than an airplane, more of the Earth's surface is captured in each photograph. A satellite also returns to the same vantage point above the Earth on a regular basis. This typically means it may be cheaper to acquire photography for larger areas or multiple sets of photography for the same area from a satellite platform than from an aerial platform.

There is a flip side to this coin. It can be very expensive, or even impossible, to obtain large-scale photography from satellites. The same challenge can arise when trying to obtain satellite photography of a specific site. The first limitation is a consequence of the camera's distance from the Earth's surface. The second limitation is a consequence of the challenges to changing a satellite's orbit. A third limitation on satellite photography is the amount of atmospheric layers a satellite camera must "see through" to capture an image of the ground surface. This limitation can reduce the resolution of the captured image. The typical aerial camera mounted on an airplane has less of the atmosphere to "see through" to capture the image.

## V. The Land Surveyor's Role in Aerial Photography

In many states, the photogrammetrist responsible for supervising the acquisition and processing of aerial photographs, as well as the production of topographic mapping from those photos, must be a licensed surveyor. California is an example of one jurisdiction that requires photogrammetrists to hold a land surveyor license. (Technical staff can assist in the mapping process if they are supervised by a licensed land surveyor.)

Land surveyors who are not photogrammetrists still have an important role in the process of acquiring aerial photos and the production of mapping products from those photos. Their role can be grouped into four primary activities.

The first activity is determining when to use aerial photogrammetry for a project and planning for that acquisition. Aerial photogrammetry is just one tool in a land surveyor's tool box. In many cases, aerial photography is the most effective and least expensive way to produce the needed data for the project. A good land surveyor will recognize these cases and assist the project team with the acquisition of the aerial photography.

The second activity is establishing the survey control for the acquisition of aerial photography and topographic mapping. The requirements for this control were explained in the previous section of this article that described the process for acquiring aerial photography for a project.

The third activity is providing quality assurance for the acquired aerial photography and topographic mapping. A land surveyor can execute ground surveys or perform other data analysis to ensure the aerial topographic mapping for a project meets the required standards.

The final activity is the supervision and execution of ground surveys to supplement the information produced from aerial photography. Ground surveys are commonly needed in areas of dense vegetation, shadows, underwater surfaces, and in the blind sides or underneath structures in order to fill in the details observed in the aerial photograph. Such surveys can also produce more accurate survey data on critical features when needed. One example would be a survey of an edge of pavement of an existing road or parking lot at the junction of a proposed road.

## VI. Conclusion

Aerial photography is a powerful tool for providing visual information and mapping data of a project site. It also has several uses in the land development and land management processes. The acquisition and processing of aerial photography and aerial topographic mapping is a multistep process that involves specialized equipment. Satellite photography is similar to aerial photography, but there are important differences between the two that should be understood before deciding to purchase satellite imaging products. All satellite and aerial photography has limitations that should be considered before its use to meet the mapping needs of a project. Land surveyors play an important role in deciding when to use aerial photography for a project and also in the process of acquiring that data.

# National Flood Insurance Program

## Floodplain Management and Regulatory Basics for Land Development

22

Wendy Lathrop, PLS, CFM

## I. Abstract

Flood loss reduction is dependant upon the ability to plan and implement appropriate floodplain management. The National Flood Insurance Program (NFIP) forms the basis of our nation's approach to floodplain management, and we must understand the program's three main areas—mapping, regulation, and insurance—in order to protect our clients, our communities, and ourselves.

Beginning with the distinctions between the various types of floods and floodplains, this article then investigates the regulatory basis for floodplain management. Understanding the terms and applicable regulations regarding floodplain management allows us to make better decisions and to advise clients regarding floodplain development plans.

## II. Introduction

Since the dawn of humankind, we have always lived and worked near water, which initially merely provided sustenance as an essential element for survival, as well as the means for bathing. Fishing first by hand and later from boats also meant proximity to water translated into less time spent traveling each day in search of food. As humans became more settled, agriculture required water to assure the success of crops. Navigation, whether for commerce or military maneuvers, formed the core of communities along the shores of rivers and oceans. Now, in more affluent areas, the pure aesthetics of living along the waterfront edges prices of building lots ever higher.

But wherever people and water interact, there is a likelihood of harm to man-made "improvements." By the very nature of gravity, water flows to the bottom of the watershed, with our rivers and streams collecting whatever water cannot infiltrate the soil or is not diverted from its normal propensity to run downhill. We may try to ignore the fact that mere scientific rules can control us, preferring to think of ourselves as self-controlling. But in the end, nature will win. Our best line of defense is to familiarize ourselves with the variables involved in flooding and to modify our own behavior to minimize the risk of flood damage.

## III. Definitions of Basic Terminology

Any time that we experience water in a place that we don't want it, we label this event a "flood." If we forget that the kitchen faucet is turned on, the overflow from the sink has "flooded" the floor. In the technical world, we define "flood" as an overflow of water onto lands that are used or usable by man and not normally covered by water. Natural floods have two essential characteristics: (1) the inundation of land is temporary, and (2) the land is adjacent to and inundated by overflow from a river, stream, lake, or ocean. Because any surplus amount of water beyond our experiences can be perceived as a flood, for regulatory purposes, a particular probability of flood risk has been established as the basis for land use regulation and calculation of flood insurance premiums.

In the United States, that particular probable flood is the 1%-annual chance flood event, sometimes referred to as the 100-year flood. Because of popular misunderstanding of "100-year flood" as only possible once per century, "1%-annual chance probability" more accurately represents the 1 in 100 or 1 percent chance of a flood of this magnitude being equaled or exceeded in any given year, and more than one such flood can occur in any year. Floods with other frequencies are calculated for land development and public safety reasons, such as the 500-year or 0.2%-annual chance flood (having a 1 in 500 or 0.2 percent chance of being equaled or exceeded in any year), the 50-year or 2%-annual chance flood, and the 10-year or 10%-annual chance flood event.

The horizontal extent of floodwaters is called the floodplain, generally an area of relatively flat and normally dry land along rivers or streams (or along lakes or ocean coasts in states where they exist) that is covered by water during a flood. Obvious floodplains are visible along beaches of the ocean and areas along shallow rivers and streams that often overflow their banks. Less obvious are the floodplains that are only covered by water under extreme conditions, such as during hurricanes or after sudden snowmelts.

Floodplains can also be areas with poor drainage that causes water to stay in an area long enough to be a nuisance, such as streets with undersized storm sewer pipes or too few inlets to carry away rainwater. Floodplains are also referred to as flood hazard areas or flood-prone areas. In the United States, the regulatory floodplain is the area that has been or may be expected to be inundated by floodwaters in a 1%-annual chance frequency flood. This area has also been referred to by the Federal Emergency Management Agency (the primary federal regulator of floodplains in the United States) as the Special Flood Hazard Area (SFHA).

The vertical extent of floodwaters is termed the flood stage, which is defined by the elevation at which overflow of the natural banks of a stream or body of water begins in the stream reach or area in which the elevation is measured. Within the 1%-annual chance floodplain, the flood stage is termed the Base Flood Elevation (BFE), being the height reached by water in a 1%-annual chance flood (the "base" flood for regulatory purposes). Water may rise above this level during a flood, but this is the base elevation at which water is considered to have reached the 1%-annual chance flood stage.

Within the floodplain may be an area known as the "floodway," although not every 1%-annual chance floodplain has a defined floodway, due to physical site factors such as topography or the presence of tidal influences. The floodway carries the bulk of floodwaters, meaning it is endangered by the greatest velocity as well as by the greatest depth and volume of water during the 1%-annual chance flood. The floodway is delineated so that even if all the rest of the floodplain is filled and unavailable to carry floodwater, there would still be capacity in the floodway to convey that water. For these reasons, the floodway is subject to the most stringent controls to prevent damage both within the rest of the 1%-annual chance floodplain and in areas beyond the limits of the 1%-annual chance floodplain to which floodwaters might extend if obstructed within the floodway. No new development may additionally obstruct or impede the flood. The balance of a floodplain containing a floodway but not affected by the floodway is termed the "flood fringe."

## IV. Types of Flooding Hazards

There are three main categories of floods:

1. Riverine flooding—Overbank flooding, the most common type of flooding in our nation, is a result of downstream channels receiving more water

than the watercourse can accommodate, so that the excess flows over its banks into the adjacent floodplain. Overbank flooding can occur from rain, snowmelt, or channel blockage by ice or debris. The terrain of the watershed will affect the behavior of the water in this type of flooding. In flat areas, shallow, slow-moving water may cover the land for extensive periods of time, while in steeper areas, the water may move so fast that it creates a flash flood. Flash floods occur as a result of sudden and intense rainfall, when the watercourses and pervious surfaces in a watershed cannot accommodate the sudden additional volume of water. While short in duration, they are often deadly.

2. Coastal flooding—Areas along the coast of oceans, large lakes (particularly the Great Lakes), and the Gulf of Mexico can experience flooding hazards not experienced in riverine areas: coastal storms and coastal erosion. Along the Pacific and Caribbean coast, tsunamis are also a hazard. The persistent high winds and air pressure changes of coastal storms push water toward the shore, causing storm surges that can raise the water level by several feet. The wind and waves shape the landforms that normally buffer areas from storms. Sand size and density, water level changes, wave effects, tides, and wind all affect whether a shore will recede or accrete. Human intervention can further erode the coastline by altering the natural systems that transport sand.
3. Shallow flooding—Areas with little change in topography are subject to shallow flooding, causing extensive land coverage by water even when there is only minimal rise in water elevation. The lack of channels in these areas can cause sheet flow, in which water flows across the land surface at a relatively uniform depth. Where rainfall collects in depressions and cannot drain, ponding is a temporary flood condition, alleviated only by infiltration, evaporation, or mechanical pumping to remove the water. Urban drainage systems are a third category of shallow flooding, as the ditches and stormwater management facilities may be overloaded by storms greater than the 10%-annual chance frequency (formerly referred to as 10-year) flooding for which the systems were designed. Increasing impervious surfaces in a watershed can compound this problem. Areas protected by levees are particularly subject to urban drainage problems, as these flat areas do not drain naturally (levees block water from flowing down to the rivers) and must be pumped. Depending upon the severity of rainfall, the urban drainage problems usually categorized as "shallow flooding" may be of much greater depths due to failure of the levee and dam systems, such as after Hurricane Katrina.

There are a variety of other 1%-annual chance flood hazards which will not be covered here but do pose significant dangers to lives and property if overlooked:

- Closed basin lakes—water bodies having no outlet, such as glacial lakes.
- Uncertain flow paths, characterized by two main examples: (1) Alluvial fans are gently sloping, fan-shaped landforms on unstable soils formed by the deposit of eroded sediment. They occur in generally arid or semi-arid regions, but can flood suddenly at extremely high speeds, carrying tremendous amounts of sediment and debris along with floodwaters. (2) Migrating stream beds move across a general floodplain with a gradual and more predictable pattern of erosion and deposition of sediment.
- Dam breaks—can result from foundation failure and/or deficient design, construction, or materials.
- Ice jams—a rapid freeze-thaw cycle can break up ice, which then drifts and blocks flow downstream, often causing predictable flooding at bridges and other constrictions to water flow that causes overbank flooding not normally experienced in those areas.
- Mudflows—landslides of liquid mud.

## V. Distinctions between Floodplain Management, Stormwater Management, and Wetlands

Some of the immediately preceding descriptions of inundated lands may seem to illustrate wetlands. Floodplains and wetlands are not the same, although they have the same basic ingredients—a mixture of land and water—and therefore may exist in similar kinds of places. In some instances, a site may be both a floodplain and a wetland. But a site may be one and not the other. Some floodplains are wet all the time, and some wetlands may be dry much of the year, blurring the distinction.

Wetlands are defined by the federal Environmental Protection Agency's "three-prong parameter," meaning that all three factors—certain hydric soils, hydrology, and vegetation—must be present to label a site as "wetland." Floodplains, in contrast, are more easily identified by the mere presence of water after certain flooding events. Flooding may be due to the presence of wetland factors upstream but not necessarily on the site in question.

Floodplain management is a program or activities that may include both corrective and preventive measures for reducing flood damage. The Federal Emergency Management Agency's online NFIP Keyword Index defines it as "a decision-making process that aims to achieve the wise use of the nation's floodplains. 'Wise use' means both reduced flood losses and protection of the natural resources and function of floodplains."

Stormwater management is a program or activities intended to manage the flow of stormwater as it flows over the ground to reduce downstream erosion, protect water quality and aquatic habitats, and flooding that is incidental to stormwater. The Federal Emergency Management Agency's online NFIP Keyword

Index has this to say: "Stormwater is water that accumulates on land as a result of storms, and can include runoff from urban areas such as roads and roofs. Therefore stormwater management is the management of stormwater runoff, often using water retention facilities, to provide controlled release into receiving streams."

One distinction between the two is that floodplain management addresses flooding from more than just storm events and can address problems associated with saturated soils or high water tables, levee breaks, the effects of erosion on water paths, and more. In contrast, stormwater management includes control of storm sewers and other facilities that carry surface runoff, street wash, and snow melt from the land, as well as water pollution measures such as the National Pollutant Discharge Elimination System (NPDES). Sanitation and water quality are more peripheral rather than central to floodplain management, which instead focuses on protection of lives and property from and prevention of flood damage.

## VI. National Flood Insurance Program: Purpose and Objectives, Legislative History, and Development

The National Flood Insurance Program (NFIP) has its legislative roots in the damages caused by floodplain development in the United States. The NFIP's basic objective is to reduce the exposure to flood damages through the use of minimum standards for the placement and design of structures located in flood hazard areas.

As early as 1956, Congress recognized that citizens who suffered property damages from flooding could not purchase flood insurance; insurance companies would not write policies in areas where flooding was known to be problematic. The 1956 Federal Flood Insurance Act was never funded, but by 1968, emergency relief funds and low-cost loans were funded with U.S. taxes.

The National Flood Insurance Act of 1968 is codified along with later congressionally created NFIP legislation in the federal statutes.[1] The NFIP was meant to create a large national pool of insured participants, thereby keeping premiums low and making the program self-sustaining in terms of paying out when disaster did strike, minimizing congressional disaster payouts. The three main accomplishments of the 1968 Act were to establish the NFIP, to require mapping of flood-prone areas, and to make federally backed flood insurance policies available in communities that met minimum floodplain management requirements as established by the Federal Insurance Administration (FIA) under the Department of Housing and Urban Development (HUD). FIA was transferred to the newly created Federal Emergency Management Agency (FEMA) in 1979.

The Flood Disaster Protection Act of 1973 mandated the purchase of flood insurance for insurable structures within flood-prone areas as a condition of

1. 42 U.S.C. §§ 4001 et seq.

receipt of federal or federally related financing. Properties in communities not participating in the NFIP were prohibited from federal flood insurance coverage or disaster assistance; participating communities are those that agree to adopt and enforce floodplain management ordinances to reduce future flood damage.

By 1986, FEMA recognized that some communities were not enforcing their floodplain management ordinances and implemented community probation procedures. This includes suspension from the NFIP if a community allows unwise practices that would undermine the purpose of the flood protection program. If a community is suspended, federal flood insurance is not available; and when current policies expire, they cannot be renewed. Flood insurance must then be purchased from outside insurers, at premiums much higher than NFIP rates. Federal disaster assistance is also unavailable during suspension.

The National Flood Insurance Reform Act of 1994 created stronger enforcement mechanisms to ensure that lenders require flood insurance for any structures within 1%-annual chance flood areas on which the lenders make, extend, or renew loans. This flood insurance requirement was not new, but one with which some lenders were not complying for several reasons. Some were not checking if buildings were flood-prone before making loans because of the additional cost and time this would take. Some may not have required flood insurance even when a site was known to be flood-prone, because informing potential borrowers of the required additional cost of flood insurance could deter those people from completing loan transactions.

The 1994 Reform Act reached beyond insurance issues and began to address the fact that many maps in the nation had not been updated since they were first created twenty years earlier. Because FEMA cannot visit each community to check the validity of its maps, the 1994 Reform Act required FEMA to contact communities every five years to ask them to review and assess their NFIP flood mapping for update needs. While not assuring such updates would be accomplished (due to limited resources), this review did begin to prioritize mapping needs across the nation.

Because physical paper maps cannot be updated and republished as often as is necessary due to funding and technical constraints, changes to floodplains are published as Letters of Map Change, but these were not, for many years, readily available, and are not automatically included with orders for maps subject to those Letters. To improve public knowledge and access, the 1994 Reform Act required publication twice per year in the Federal Register of a compendium listing the Letters and other flood mapping changes. Over time, these Letters have been made available within weeks of issuance through FEMA's Map Service Center on the Internet.

Acknowledging the constant changes caused by the interaction of soil and water, the 1994 Act also mandated evaluation of coastal and riverine erosion hazards and the effects of erosion on flooding (these studies were accomplished in 1999). Additionally, the 1994 Act established a five-year Technical Mapping Advisory Council to FEMA to provide recommendations on improving every aspect of flood mapping, including data gathering, data analysis, data reporting, and data distribution.

Subsequent legislation continues to improve enforced protective measures while making the program more financially sound. The Flood Insurance Reform Act of 2004 implemented the first measures to cease payments for damages to structures that were subject to repetitive losses due to flooding. At the same time, funding was increased to mitigate repetitively damaged structures by moving them or elevating them. Recognizing that the digital age had arrived and was technically sound as well as financially beneficial, the 2004 Reform Act also acknowledged certain digital versions of FEMA's paper maps as being legally equivalent. As of October 2009, paper maps are no longer distributed beyond a single set to participating communities; all other map copies are digital and available through FEMA's Map Service Center.

## VII. The Three Facets of the NFIP

The NFIP basically works through three approaches to protecting against and minimizing flood damage: mapping, insurance, and regulation.

- Mapping—A study of the area results in a map that is used to determine if structures are within the 1%-annual chance floodplain, assess insurance premiums, and determine where floodplain management regulations must be enforced. Without a map, the public has no notice of where the floodplains are.
- Insurance—Flood insurance rates are established based on whether the construction preexisted the mapping of the floodplain (pre-FIRM) or occurred after the publication of a map identifying the floodplain (post-FIRM), the type of construction, and the height above the Base Flood Elevation. Individual properties not properly covered by insurance but located in an identified 1%-annual chance floodplain are not eligible for disaster assistance.
- Regulations—FEMA sets national policy for floodplain regulation and administers the flood hazard mapping program. State agencies act as the local coordinators for the NFIP, and local communities must adopt their own ordinances meeting minimum national criteria. Thus, while the federal government establishes minimum floodplain management standards, the community has the strongest role in sound floodplain management.

The federal regulations serving as the basis for the National Flood Insurance Program are found in 44 C.F.R. §§ 59–78. These regulations describe the program, define the terms used to run the program, provide minimum floodplain management criteria for communities to adopt and enforce, provide technical criteria and requirements for revising and amending the flood hazard areas on the maps, and codify fees charged for reviewing map changes.

Mapping, insurance, and regulation are intertwined throughout the NFIP. Without current and accurate mapping of flood hazards, properties can be either underinsured or overinsured, and underregulated or overregulated. Without regulation, construction in floodplains can put properties at unnecessary risk, meaning that insurance rates (which are based on risk) increase.

## VIII. NFIP Regulations and Mapping: Significance to Communities and Lenders

In order to properly implement and enforce regulations, floodplain mapping must be in place, but federal regulations create a process for communities to join the NFIP even before mapping is completed, to allow residents and business owners the opportunity to buy federal flood insurance (which is not available in communities not participating in the NFIP but is significantly less expensive than other such insurance policies).

When a community first joins the NFIP, it is given "emergency" phase status. During emergency phase, detailed maps identifying flood-prone areas are not yet prepared. In joining the NFIP, a community agrees to certain minimum floodplain management practices and must adopt ordinances to enforce those practices to protect its floodplains. Because of the community's demonstrated commitment to joining the program (by adopting ordinances complying with minimum federal standards), residents and business owners are eligible to purchase flood insurance from the NFIP.

A community remains in the "emergency" phase until mapping of the flood zones is completed and accepted by the community, at which point it enters the "regular" phase of the NFIP. In the "regular" phase of participation, maps and regulations are more detailed than in the "emergency" phase, and flood insurance premiums are generally lower.

Communities falling out of compliance with the minimum NFIP regulatory requirements, which can happen by not enforcing its floodplain management ordinances or not adopting newer flood hazard mapping, are subject to suspension and possible expulsion from the NFIP. In either of these circumstances, the community is treated as a nonparticipating community: existing federal flood insurance policies cannot be renewed, and no new federal flood insurance policies can be written. Further, in communities expelled from the NFIP, no new federal insurance coverage is available, and the community is ineligible for federal disaster assistance. This serves as strong impetus to follow the regulations closely, and communities usually work hard to maintain compliance with them.

A federal incentive program encourages communities to exceed the minimum responsibilities defined by 44 C.F.R. § 60. The Community Rating System (CRS) rewards communities that enforce proactive floodplain management, land use, and construction regulations by crediting such communities with points that can result in significant reductions in flood insurance premiums for every policyholder within that local jurisdiction.

## IX. Mapping Floodplains

Federal regulations uniformly establish the 1%-annual chance flood as the trigger for NFIP mandates at every level of governmental control or oversight. FEMA issues maps identifying 1%-annual chance floodplains called Flood Insurance Rate Maps or FIRMs, an increasing percentage of which were created and are distributed in digital form and are therefore abbreviated as DFIRMs (Digital Flood Insurance Rate Maps).

The dangers and risk of structural damage in the floodway are greater than within the flood fringes, and therefore regulations restrict the allowable land uses and types of construction beyond general 1%-annual chance floodplain restrictions. The depth, speed, and volume of water within the floodway mean that evacuation is even more difficult than within the flood fringes, and so land use regulations generally prohibit certain facilities in floodways, particularly hospitals, schools, and prisons.

## X. Lending Institutions and Flood Insurance

For structures located within the mapped 1%-annual chance floodplain, flood insurance is mandatory for every loan issued by lenders regulated by a federal agency (banks, savings and loans, credit unions, thrifts, Fannie Mae, Freddie Mac); unimproved land is not insured against flood damages. Lenders must verify the flood hazard status of structures on collateral property for loans of every type, whether issuing a new loan, renewing a loan, extending or refinancing a mortgage, or issuing a home equity loan.

If borrowers refuse to purchase flood insurance on their own, then lenders can force place it by purchasing it on behalf of the borrower and including the cost in escrow accounts. Lenders must also ensure that borrowers maintain flood coverage for the life of the loan, which means that lenders must monitor the FIRMs for changes that may either remove structures from the mapped floodplains or newly place them in the 1%-annual chance floodplain and therefore make them subject to mandatory flood insurance purchase requirements. When loans are transferred between lending institutions, the requirement of "life of loan" flood insurance coverage travels along with those loans. The record-keeping process involves the use of the FEMA-issued Special Flood Hazard Determination Form (or its electronic equivalent) to standardize the information and format in which it is kept. This simplifies the auditing process for ensuring that lenders are enforcing mandatory flood insurance coverage when federal laws require it.

A large number of flood hazard areas have not been delineated, yet they present a serious threat to people who may choose to buy or build within them. The most pressing problems exist in or near communities that are growing, but this is not the only area of concern. Although not all locally known flood-prone areas are identified on FEMA's mapping, such areas may still be regulated locally, and lenders may still require flood insurance as protection of their collateral, even though federal regulations do not seem to apply.

## XI. Distinctions between Flood Zones

44 C.F.R. § 59.1 establishes the basis for our nation's flood protection as the "area of special flood hazard," or "the land in the flood plain within a community subject to a 1 percent or greater chance of flooding in any given year." While this 1%-annual chance floodplain is the most regulated area of risk, every area on the surface of the earth has some risk of flooding, whether 1%-annual chance risk, moderate risk (0.2%-annual chance), or minimal risk (less than 0.2% chance). No area is identified on FIRMs or DFIRMs as having "no risk."

The symbols, or lettered designations, shown on the NFIP maps to represent the 1%-annual chance and less probable flood risks are established in 44 C.F.R. § 64.3, and include zone designations for areas of 1%-annual chance flood hazards, coastal high-hazard areas, flood-related erosion areas, and flood-related mudflow or mudslide areas. The majority of 1%-annual chance floodplains are either Zone A or Zone V types, with the distinction between them being that the V-type zones are additionally subject to high-velocity wave action and surge from storms or seismic sources. V-type zones are found along coastal areas of the Atlantic and Pacific Oceans, the Gulf of Mexico, and the Great Lakes. A-type zones may be found along the coastlines where storm surge is not a factor, but are usually associated with riverine flooding.

Within both A and V types of zones, there are areas shown only approximately, meaning that no Base Flood Elevation (BFE) has been determined for them. Within both A and V types of zones, there are also areas of shallow flooding for which only a depth of flooding is determined to the nearest whole foot, rather than a BFE per se. The determination of flood risk for any given site within either an approximate or shallow flood hazard area is therefore less consistent and less accurate than in areas where a BFE is provided. Local communities may require additional freeboard (a safety factor expressed in feet) above the approximate height of the base flood in order to provide additional protection in these areas of uncertainty. Communities can also require freeboard in areas where BFEs are established, also as a means of additional protection, as the height of the predicted BFE reflects a 1%-annual chance probability or risk that can be exceeded.

Identifying the zone in which a given structure is located is integral to the functions of the insurance and land use regulatory aspects of the NFIP:

- Lenders must verify flood zone designations for structures serving as collateral on loans. For any such structures located within a regulated 1%-annual chance floodplain, government-regulated lenders are legally required to impose the mandatory flood insurance purchase requirements that were first introduced by the 1973 Flood Disaster Protection Act.
- Flood insurance rates are based upon the level of risk to a structure as first identified by the mapped flood risk zone and then, if within the 1%-annual chance floodplain, by how far above or below the Base Flood Elevation the lowest floor or lowest structural member of the building is. Less risk means a lower annual premium.
- Mapped flood zones guide community land use as well. By their participation in the NFIP, communities agree to regulate all land use and development in the 1%-annual chance floodplain and to impose a permitting process that acknowledges the contribution of runoff from non-1%-annual chance floodplain portions of a watershed to the risks at the bottom of the hill where the highest risk floodplains are located.

## XII. Insurance Issues: When Insurance Is Required, Defining Lowest Floor for Insurance Purposes

NFIP regulations require structures within the 1%-annual chance floodplain to be covered by flood insurance only if they secure a loan; owners who are able to purchase without a mortgage are not subject to the mandatory purchase requirement because they do not interact with the lending institutions serving as the gatekeepers for the NFIP. Separate policies exist for structural coverage and for contents, a point particularly significant for condominium owners. The owner of a unit on the tenth floor of a structure within a mapped 1%-annual chance floodplain must participate in a homeowners' association flood insurance policy for the structure, because the common elements and even access to the building itself are part of the unit owner's insurable interests. Individual unit owner's premiums for contents will be less on the tenth floor than on the first, however, because the risk of flood damage to those contents is less.

There are a few exemptions from the mandatory flood insurance coverage requirement. One is for government-owned land if the owner is self-insured. Another exemption is for situations in which the original loan was for an amount less than $5,000 and duration of one year or less.

There are also exclusions from coverage, meaning structures that are ineligible for federal flood insurance. These include buildings that are in violation of state or local laws and buildings (or boathouses) erected entirely over water after October 1, 1982. Other structures ineligible for federal coverage include those known to be at risk of repetitive damage from flooding, with congressional legislation increasingly tightening the amount of damage (both in terms of dollars and in terms of the percentage of the structure's value) to prevent repetitive payouts. Early restrictions on the sale of federal flood insurance went into effect with the Coastal Barrier Resources Act (CBRA), prohibiting coverage in identified areas for structures completed after or substantially competed by November 16, 1990. A second prohibition applies in Otherwise Protected Areas (OPAs) for structures completed after or substantially competed by November 16, 1991.

Insurance premiums are rated differently between the two types of 1%-annual chance floodplains. In A-type zones, the elevation of the lowest floor of the structure, including the basement, is compared to the Base Flood Elevation (BFE) to determine the risk to the overall building. If the lowest floor is at or above the BFE, rates are considerably less expensive than if the lowest floor is below BFE. The rate increases as the lowest floor sinks below BFE and decreases as the lowest floor rises above BFE. Within areas subject to storm surge in V-type zones, it is not the lowest floor of the structure but the lowest structural member of the building that is compared to the BFE to determine rates. For instance, a home on pilings must have its lowest crossbeam at or above BFE for lowest premiums.

In both A- and V-type zones, any machinery or equipment that services the building must also be elevated for the greatest reduction in insurance rates. In V-type zones, this requires elevating air handlers and other equipment on platforms at or above the level of the lowest structural member. Enclosures beneath buildings in V-type zones must not be greater than 300 square feet and may only be used for parking, storage, or access to upper levels. Failure to observe these requirements results in substantially higher premiums because obstructing the flow of water beneath an elevated building means that wave impact and water pressures are increased. FEMA's Technical Bulletin 5, "Free of Obstruction Requirements," elaborates on these requirements. Technical Bulletin 9, "Design and Construction Guidance for Breakaway Walls Below Elevated Coastal Buildings," provides a good supplement.

In A-type zones, residential buildings must be vented or have foundation openings to allow the pressure of floodwaters to equalize on both sides of the wall. While owners may not like the idea of floodwaters entering their homes, the structural damage to a building that fails to withstand water pressure may be greater than the damage of silt in the basement. Appropriate flood openings allow the flow of water into the building without any human intervention: windows, doors, and air vents do not qualify as appropriate flood openings. Buildings in A-type zones must have a minimum of two flood openings per enclosure, on different walls, with a minimum of 1 square inch of opening for every enclosed square foot. The bottom of flood openings must be within 12 inches of the adjacent grade. If the lowest floor of the structure is built so that water cannot flow from one part to another without human intervention of opening a door or window, then each separate enclosed area must meet these requirements. Proper flood openings allow insurance agents to write policies based upon the next higher floor. Improperly flood-vented residential structures are rated according to the lowest floor elevation, significantly increasing flood insurance premiums. See FEMA's Technical Bulletin 1, "Openings in Foundation Walls," for more detail.

There is also a difference in flood insurance rates for structures completed before an area has been mapped as being in the 1%-annual chance floodplain on the Flood Insurance Rate Map (pre-FIRM) and structures completed after an area has been mapped as subject to the 1%-annual chance floodplain (post-FIRM). The rationale for this is that in the first instance, the building was presumably erected in good faith that it would not be at risk, while in the second, a public document (the FIRM) provided notice that there was a risk of flooding in that location. This distinction may eventually evaporate in terms of premiums as our nation experiences greater frequency of flooding disasters, in an attempt to more closely reflect the cost of true risk for all insured properties. While the NFIP was designed to be a self-supporting program that would pay for claims from premiums collected, the number of flooding disasters and the extent of damage has resulted in expenses that far exceed what the policies bring in.

## XIII. Various Approaches to Floodplain Management

Floodplain management includes both structural and nonstructural approaches. Structural approaches are, as the name implies, based upon human construction of flood-control works such as dams, levees, sea walls, and other facilities to protect from flood risks and to control floodwater or at least provide a window of time for evacuation. Structural projects can include reservoirs and detention basins to hold excess water upstream, dikes and floodwalls to keep water away from developed areas, and altering stream channels to make overbank flooding less frequent.

Since the earliest days of this country, human response to water has been to try to control it in order to "reclaim" land. Protection against flooding was generally limited to structural methods of floodplain management. During the twentieth century, this method was supplemented by providing disaster relief to flood victims. These methods neither reduced losses nor discouraged unwise development. In some instances, they may actually have encouraged additional development. Meanwhile, flood coverage from insurance companies was either exorbitantly expensive or not available, and building construction rarely included techniques to reduce flood damage.

In contrast, nonstructural approaches are meant to guide human behavior and generally take the form of floodplain management regulations. To modify human susceptibility to flood damage by avoiding hazardous or unwise use of floodplains, nonstructural means include zoning ordinances, subdivision and land development regulations, building codes, health regulations, land acquisition and preservation ordinances, public education about floodplain purposes, and emergency preparedness.

Nonstructural approaches may reap benefits beyond reducing flood risks. For example, creating riparian buffers and leaving floodplains undeveloped (or returning them to that status through buyouts of repetitively flooded structures) provides passive recreational open space and wildlife habitat while protecting water quality by reducing erosional forces of rapid flow over impervious surfaces. Though we may never be able to avoid building near water, we should recognize that the cumulative effects of construction anywhere within the watershed increase both volume and velocity of water in the lower reaches where the floodplains lie.

We can manage but cannot control flooding. Instead, avoiding a problem can lessen or even prevent later loss and damage; this approach is called mitigation. FEMA has estimated that for every dollar spent in flood damage prevention, two are saved in repairs. In the late 1990s, FEMA's Mitigation Directorate posted the following description on its Mitigation Planning website:

> To achieve the goals of floodplain management, the nation must adopt a new approach—one that takes full advantage of all methods available to reduce vulnerabilities to damages and, in parallel, to protect and enhance the natural resources and functions of the floodplain. This approach would achieve floodplain management through:
>
> - avoiding the risks of the floodplain,
> - minimizing the impacts of those risks when they cannot be avoided;
> - mitigating the impacts of damages when they occur; and
> - accomplishing the above in a manner that concurrently protects and enhances the natural environment.

A significant aspect of mitigation and flood awareness lies in planning for the worst and in learning from the experiences of natural disaster. Sound floodplain management includes adopting and enforcing local land use regulations, planning for disaster warning and rapid response systems, evacuation routing, and post-disaster verification of actual flood map accuracy in predicting the areas and extent of flooding. FEMA and other agencies increasingly pursue mitigation and have established various programs in recognition of its value in preventing damages.

## XIV. NFIP Regulations: Land Use Management Aspects (44 C.F.R. § 60)

There are two purposes of protective measures in floodplains; one is to protect human life and property, and the second addresses water quality. In the United States, FEMA administers the NFIP to reduce risks due to flooding. Land use and floodplain management requirements established at the federal level must be adopted as the minimum criteria for state and local regulations. More stringent regulations may be adopted and enforced, such as requiring greater elevation above predicted flood levels than mandated by federal regulation. There may be additional regional watershed and river commissions with jurisdictional control as well.

Our federal government has set minimum criteria for safe use of floodplains, establishing the 1%-annual chance floodplain as the trigger for enforcing floodplain management controls and NFIP mandates at every level of governmental control or oversight. Regulatory restrictions can prevent construction of any new structures or additions to existing structures in the floodplain, require floodproofing or elevation of structures and accessories, and prevent certain uses for new or existing structures (such as prohibiting hospitals in floodplains because of the problem of evacuation).

The minimum requirements for community ordinances are stated in 44 C.F.R. § 60, addressing the various types of flood risk zones present in a community and the restrictions to development, definition of allowable construction techniques and materials, and requirements for elevation of structures based on the mapped zones.

Sound use of land is a primary tool in achieving protection from flooding. Part 60 of 44 C.F.R. establishes the minimum criteria for land use regulation in communities that participate in the NFIP. Communities may always enact and enforce more stringent regulations, but they may not be more lenient. They must also be consistent in their enforcement of land use regulation, which can be difficult in the face of pressure for variances. But a pattern of nonenforcement or inconsistent enforcement of ordinances can result in the reassessment of a community's status in the NFIP, threatening the availability of federal flood insurance and disaster assistance, as well as opening the door to lawsuits by landowners in civil court.

The land use management aspects of the federal guidance provided in 44 C.F.R. § 60, entitled "Criteria for Land Management and Use," is divided into three subparts, although only portions of subpart A will be examined here.

- Subpart A—Requirements for Flood Plain Management Regulations
- Subpart B—Requirements for State Flood Plain Management Regulations
- Subpart C—Additional Considerations in Managing Flood-Prone, Mudslide (i.e., Mudflow)-Prone, and Flood-Related Erosion-Prone Areas

### *A. 44 C.F.R. § 60.1—Purpose of Subpart*

- Subpart A establishes minimum criteria for community compliance with NFIP

- Communities may adopt more comprehensive floodplain management regulations
- State and community regulations that are more stringent than federal minimum criteria are encouraged, and take precedence over NFIP minimums

### B. 44 C.F.R. § 60.2—Minimum Compliance with Flood Plain Management Criteria

- Communities shall meet Section 60.3 requirements for flood insurance eligibility
- Mudslide-prone (mudflow) properties must meet Section 60.4 requirements
- Flood-related erosion-prone properties must meet Section 60.5 requirements
- If multiple hazards exist, all applicable regulations in Sections 60.3, 60.4, and/or 60.5 must be met

### C. 44 C.F.R. § 60.3—Floodplain Management Criteria for Flood-Prone Areas

Section 60.3 provides the heart of most community regulatory programs and is comprehensive in its scope regarding development and placement of fill in 1%-annual chance floodplains.

The table below identifies which paragraphs of section 60.3 are applicable to various flood-prone conditions shown on the FIRM in effect within a community. All requirements are cumulative, so that requirements for conditions in the rows at the bottom of the table include requirements that precede them. In other words, if paragraph 60.3(d) applies, so do paragraphs 60.3 (a), (b), and (c).

The regulations within 44 C.F.R. § 60, which apply to all properties within identified 1%-annual chance floodplains, should be read in their entirety for full guidance. Highlights of 44 C.F.R. § 60 are presented after the table to explain the regulations applicable when various conditions are present. Zones AR and A99 are not discussed here but are addressed in the regulations.

**44 C.F.R. § 60.3–Flood Plain Management Criteria for Flood-Prone Areas**
**FLOODPLAIN MANAGEMENT ORDINANCES AND NFIP REGULATIONS**

| 1%-annual chance flooding | BFEs | Floodway | V Zones | Land Use Classification |
|---|---|---|---|---|
| No | No | No | No | 60.3(a) |
| Yes | No | No | No | 60.3(b) |
| Yes | Yes | No | No | 60.3(c) |
| Yes | Yes | Yes | No | 60.3(d) |
| Yes | Yes | No | Yes | 60.3(e) |
| Yes | Yes | Yes | Yes | 60.3(d) & 60.3(e) |

### *D. 44 C.F.R. § 60.3(a)*

- The community must require permits for all proposed construction or other development, including placement of manufactured homes.
- The community must review permit applications to determine if projects will be reasonably safe from flooding [44 C.F.R. § 65.2(c): 'reasonably safe from flooding' means base flood waters will not inundate the land or damage structures to be removed from the 1%-annual chance floodplain and that any subsurface waters related to the base flood will not damage existing or proposed buildings." *See* FEMA Technical Bulletin 10, "Ensuring that Structures Built on Fill In or Near Special Flood Hazard Areas are Reasonably Safe From Flooding."]

### *E. 44 C.F.R. § 60.3(b)*

- Base Flood Elevation must be established for developments of 50 lots or 5 acres, whichever is less.
- The community must keep a record of lowest floor elevations of structures in A-type zones.

### *F. 44 C.F.R. § 60.3(c)*

- Lowest floor of structures in A-type zones must be elevated to or above the Base Flood Elevation (BFE).
- Structures in A-type zones must be vented and floodproofed.
- Lowest structural member of structures in V-type zones must be elevated to or above the Base Flood Elevation.
- Cumulative effects of development are not to increase BFE by more than 1 foot (future conditions accounted for).

### *G. 44 C.F.R. § 60.3(d)*

- Floodway encroachment and development prohibited unless no rise in BFE would occur.
- Conditional Letters of Map Revision are required prior to development.

### *H. 44 C.F.R. § 60.3(e)*

- New construction must be landward of mean high tide.
- No fill is permitted in V-type zones.
- Manufactured homes must be elevated and anchored.

### *I. 44 C.F.R. § 60.4—Flood Plain Management Criteria for Mudslide (I.E., Mudflow)-Prone Areas*

- Communities must enforce a permitting and review process.
- Communities must enforce grading, drainage, and excavation regulations.

### J. 44 C.F.R. § 60.5—Flood Plain Management Criteria for Flood-Related Erosion Prone Areas

- Communities must enforce a permitting and review process.
- Relocation or protective measures for proposed structures.
- Requires setback from water courses with a vegetative or contoured buffer.

### K. 44 C.F.R. § 60.6—Variances and Exceptions

- No absolute criteria for variances are established by FEMA.
- Insurance premiums are not affected by granting of a variance; variances only exempt properties from floodplain management regulations.
- Variances are allowable for historic structures and for structures functionally dependent upon water.
- No rise in Base Flood Elevation is permitted if the site is in the floodway.
- Floodproofing and flood warning times are established by FEMA.

### L. 44 C.F.R. § 60.7—Revisions of Criteria for Flood Plain Management Regulations

- From time to time, part 60 may be revised as experience is acquired under the NFIP and new information becomes available. Communities will be given six months from the effective date of any new regulation to revise their floodplain management regulations to comply with any such changes.

While the regulations mentioned here might seem onerous, they do not prohibit construction within 1%-annual chance floodplains. There are regulatory requirements and then there are technical requirements, and engineering feats of wonder may pass the regulatory test.

But although a structure may be constructed in compliance with technical regulations, this does not mean it will be exempted from mandatory flood insurance purchase requirements. In most instances, construction that meets the letter of the law without acknowledging the spirit of the law will be assessed extremely high flood insurance rates—and such insurance may not be available from all sources. For instance, a structure can be erected with its lowest floor 3 feet below Base Flood Elevation, with 3 feet of floodproofing that reaches that BFE. While the technical aspect of floodproofing provides some protection, the location of the lowest floor so far below the expected level of flooding guarantees an exorbitant annual premium.

Letters of Map Revision Based on Fill (LOMR-Fs) are waivers of mandatory flood insurance coverage granted for sites raised above Base Flood Elevation (BFE) by the placement of fill on-site after the publication of the currently effective

Flood Insurance Rate Map (FIRM). The purpose of filling a site is to raise it above the risk of flooding. But in granting such waivers from mandatory flood insurance coverage, FEMA requires that the lowest floor of structures on filled sites be constructed above BFE, recognizing that it is not just surface water that poses a risk to structures. Infiltration and soil saturation pose risks to structures with basements and/or foundations below BFE, due to hydrostatic pressures. Therefore, a LOMR-F will not be issued for structures built on fill with lowest floor elevations below BFE. Technical Bulletin 10 provides technical guidance for building below BFE into fill, advising of the most secure foundation types. But this is only for structural soundness, not for insurance and regulatory waivers. Such buildings are still within the 1%-annual chance floodplain and must still be insured against flood damage.

Once a new FIRM is published, no longer showing the affected area within the 1%-annual chance floodplain, an area that had been considered "fill" will become a preexisting condition, and the area may no longer be regulated floodplain. However, filled soils do not reach the same compaction levels as natural soils, no matter how well constructed, and communities have several options to continue their oversight, as they are required by federal regulation (44 C.F.R. §§ 65.5, 65.6) to ensure that development is "reasonably safe from flooding." To uphold this responsibility, communities can regulate construction in filled sites by prohibiting basements by requiring deed restrictions that serve notice of the filled condition and by prohibiting floodplain fill altogether.

## XV. Concluding Remarks

The land use restrictions imposed by the presence of floodplains are both physical and regulatory. The general location of a site at the bottom of the watershed, as well as its soils, geology, and topography, contribute to the risk to lives and property from flood hazards. These are conditions that should impose far more consideration than mere adherence to floodplain management regulations and ordinances, but their significance is sometimes minimized in the pursuit of possible economic gain. Coastal, riverine, and other flooding hazards each pose their own dangers to soil stability, structural soundness, and human safety—dangers that flood hazard mapping may not communicate. Sound floodplain management and floodplain development incorporate mitigation as well as more traditional structural and nonstructural approaches and ideally achieve their goals without adversely impacting other properties in the watershed.

Technically permissible floodplain development may adhere to regulatory requirements that minimize (but do not eliminate) risks of flood damage, but flood insurance may still be mandated when a federally backed loan is involved in that development. This distinction between technical compliance and insurance requirements is too often overlooked by property owners and developers, resulting in unanticipated and expensive annual insurance premiums.

Encroachments into the 1%-annual chance floodplain can result in repercussions ranging from denial of disaster assistance for individual property owners to community suspension from the NFIP, particularly if a community is found to have a pattern of allowing such development in conflict with its own ordinances or federal minimum standards for floodplain management.

# Licensure and Responsibilities of Land Surveyors

23

Wendy Lathrop, PLS

## I. Purpose of Licensing Land Surveyors

Each of the states and territories of the United States has its own set of laws governing who may be licensed as a land surveyor and what the scope of professional practice is for the surveyor in that jurisdiction. There is, however, an underlying theme running through all of this legislation, and that is protection of the uninformed consumer or layperson. The phrase commonly repeated is "to protect the health, safety, and welfare of the public," or words generally to that effect.

What is it that the land surveyor does that must be regulated for such grand humanitarian purposes? The territorial instincts of humankind and the historically resultant bloodshed are one indicator of the immensity of the responsibility for properly locating lines of ownership and possession on the ground. In more recent history, the growing list of the surveyor's duties include, among other things, the careful staking out of construction in accordance with land use regulations (such as setback lines, or how close structures may be to property lines), notification of and respect for easement rights (so that rights given to one party over the land of another, such as for private access or for maintenance of a public utility line, are not obstructed), and creation of highly accurate base mapping for geographic information systems.

As technology grows, the charges of the surveyor in relation to land issues multiply accordingly. Another source of the growth may be the increasing trend toward litigation and the resulting need for expanding reliance upon the services of an expert licensed individual.

## II. Definitions of Land Surveying Practice

The actual definition of allowable practice, or scope of services permitted by regulation, varies somewhat between the states. Some of this variation arises from the basic historical growth of land surveying away from the practice of civil engineering and how far this separation has evolved in different areas.

Some jurisdictions define land surveying as a branch of civil engineering and therefore include certain land-related design functions in the practice of surveying. This generally is restricted to small-scale lot grading up to full drainage system design, but may include road design or other functions typically recognized as being engineering. Other states prohibit any small foray into the realm of the engineer and clearly define land surveyors as those who must be expertly trained in measurements and knowing the law pertaining to land and/or property boundaries.

In some states, engineers are permitted to do certain topographical surveys, meaning measurements to determine the contours of the earth's surface, as they pertain to design purposes. These rights are generally restricted to be strictly interior to a parcel of land so that determination of the boundaries of the site is still solely under the purview of the land surveyor, and the plans prepared by the engineer often may not relate the topographic data to the boundaries of a parcel without the signature of a licensed land surveyor. However, if an engineer is also licensed as a surveyor, a single signature by that individual will obviously suffice.

## III. Acquiring Professional Licensure

To be licensed as a land surveyor, the professional must meet certain minimum criteria set by the individual state Boards of Registration. These requirements may consist entirely of years of experience acceptable to the board (such as showing an increase in responsibility, a suitable range of experience in types of surveys performed, especially in land boundary surveying, and a certain ratio of field-to-office experience), a combination of experience and education (the latter may range from a two-year to four-year degree requirement), or possession of an out-of-state license. If that license was obtained under criteria similar to that of the board's to which application is being made, the board may accept the foreign license as reciprocal and make no further requirements, or some portion of the formal examination may be required, such as the state's own jurisdictional exam.

The current examination system accepted by most of the states comprises fourteen hours of testing compiled and administered by the National Council of Examining Engineers and Surveyors (NCEES), containing sections on the fundamentals of surveying (eight hours) and the principles of practice (six hours). These standardized tests are supplemented by each state's own two-hour (or more) exam on the aspects of surveying practice deemed most significant by the individual state board. Thus the state-administered portion of the exam may consist of two hours of ethics and law, or eight hours of drainage and road design.

The common denominator for current licensing examinations is a sixteen-hour minimum, but this is only since a transition during the 1980s. Earlier licensees may have had an oral interview, or an eight-hour mathematical exam, or no exam at all—just the testimony of other licensees as to the applicant's qualifications to obtain professional status. Some licenses were granted by "grandfathering," based on years of practice prior to licensing regulations. For a while, some states may have granted land surveying licenses to engineers upon application to the licensing board for exemption from examination requirements.

## IV. Exemptions from Licensing Requirements

There are exceptions from the licensing requirement. One such case is for government employees when their work is restricted to work solely for their employer. However, any outside, nongovernment work must be supported by a professional license. Certain construction layout work is also exempt from licensing requirements, but generally only when a licensed surveyor has previously established boundary, horizontal, and vertical control. Employees of utilities and railroads have an exemption similar to that of government employees. Generally all other unlicensed surveyors must be under the responsible charge and supervision of a licensee to perform land surveying services, especially services provided to the general public.

## V. Maintenance of Professional License

A new trend in license maintenance is the requirement for mandatory continuing education or professional competency in many states. This practice is becoming more prevalent as the boards realize that the technical and legal aspects of the surveying profession do not remain stagnant over time but do evolve with technology and the outcome of legislation and litigation.

In the states having some variation on this requirement, a certain number of hours of education or professional activities are required before the renewed license is issued. Some states require twenty-four contact hours of classroom experience every two years; others may require thirty hours for the same period or allow active participation in a professional society, writing and presenting papers, or classroom instruction to count toward the renewal requirements.

Generally, those individuals not meeting the hours or quality of experience set by the licensing board are denied license renewal; it is not a matter of simply paying a fine. However, the review of CEU (continuing education units) or PDH (professional development hours) may be on a random audit basis so that not every licensee is required to prove the validity of credits claimed.

## VI. Revocation and Suspension of License

A surveying license may be suspended or revoked by a licensing board for noncompliance with regulations. This may include insufficient CEU hours or conviction on a matter of misconduct. Reinstatement of a suspended license may involve probationary time, monetary fee, and additional classroom attendance.

Revocation of a license is reserved for the most flagrant of violations or repeat offenses, and may prohibit the individual from ever reapplying for licensure in that state again. Some boards may require entirely new application and references before considering moral suitability for new examination.

## VII. License Designations

Most states have only one designation for licensed individuals. Each state will designate the appropriate title and abbreviation as PLS (Professional Land Surveyor), CLS (Certified Land Surveyor), RLS (Registered Land Surveyor), LLS (Licensed Land Surveyor), PS (Professional Surveyor), or RPLS (Registered Professional Land Surveyor).

A few jurisdictions do distinguish between classifications of licensure and thereby restrict the practice of certain license holders. For example, in Maryland, the Property Line Surveyor (with no abbreviation allowed in order to avoid confusion) is only permitted to deal with boundary matters, while the Professional Land Surveyor (which is also spelled out), is also licensed to perform certain drainage and road design functions. Though those holding Property Line Surveyor licenses may renew them, this category of license is no longer issued, and all newly licensed surveyors must qualify as Professional Land Surveyors.

A few states, such as Virginia and North Carolina, formerly issued A and B or Tier I and Tier II licenses, dividing the permissible scope of services between boundary work only and the addition of design aspects. Although no longer issued, New York formerly granted "E" exemptions to allow some land surveyors to do limited engineering design work.

Additionally, a growing number of states provide for an interim status between the unlicensed practitioner and fully vested licensed professional. This is the

LSIT, or Land Surveyor in Training. The holder of this title must have certain experiential and education background before sitting for the eight-hour Fundamentals portion of the licensing examination.

Successful LSITs are required to obtain additional years of experience before being permitted to sit for the six-hour Principles and Practice section and the state-specific portion of the exam, which they must do in order to obtain a full land surveying license. An LSIT may not sign or seal work products, but the status represents the achievement of a certain level of professional expertise on the way to full licensure. The LSIT has demonstrated sufficient qualifications to be permitted to take, and has passed, half of the land surveyor's examination.

## VIII. Conclusion

In contracting the services of a land surveyor, it is wise to clarify the scope of services required in a particular locality to be sure that the land surveyor is the appropriate professional to perform the work. Each state may slightly redefine the allowable practice of land surveying, which may include some minor engineering activities in some jurisdictions, while those same activities are strictly prohibited in others.

The Board of Registration in each state issues regulations defining the practice of surveying and receives complaints regarding poor practice or misconduct of persons licensed under their jurisdiction. Each state will define the appropriate title for the licensed professional, and some may also provide for the Land Surveyor in Training as an interim in the licensing process.

### *Selected References*

DEL. CODE ANN. tit. 24, §§ 2701–2706.
FLA. STAT. §§ 472.001 et seq.
MD. CODE ANN., BUS. OCC. & PROF. §§15–101 et seq.
N.Y. EDUC. LAW § 7203.
N.C. GEN. STAT. §§ 89C-1 et seq.
OR. REV. STAT. §§ 672.005 et seq.
63 PA. CONS. STAT. §§ 148 et seq.
TEX. OCC. CODE ANN. §§ 1071.001 et seq.
VA. CODE ANN. §§ 54.1–400 et seq.

# Ethical Considerations for the Professional Surveyor[1]

24

Robert W. Foster, PE, PLS

One definition of ethics in the context of surveying is "the possession of appropriate personal and professional values and behaviors and the ability to make sound judgments when confronted with ethical dilemmas in a professional context."[2] A confrontation with ethical dilemmas occasionally requires courage on the part of the professional, and surveyors are confronted with ethical dilemmas more often than they may realize. One of those dilemmas is the ubiquitous conflict of interest, or the even more insidious *appearance* of conflict of interest. Another is a question of confidentiality.

## I. Conflict of Interest

We are familiar enough with the conflict of interest dilemma faced by public officials, and there are many examples in the news of public figures being accused and convicted of conflicts of interest. The professional in private practice who also serves in an official position for a local government may be in a precarious ethical position. To their credit, many surveyors contribute their time and energy to their communities as elected or appointed members of planning commissions, boards of appeals, conservation commissions, and so on. For them, the conflict will come (sooner or later) when a project appears before

1. An earlier version of this article appeared in *P.O.B.*, March 2005.

2. Stig Enemark & Frances Plimmer, Mutual Recognition of Professional Qualifications in the Surveying Profession, Proceedings of the FIG Working Week, Prague (2001).

the board in which they have had some past connection. The connection need not be professional—a family connection is enough to create a conflict. The issue becomes a dilemma when the connection is distant and obscure. The professional may honestly believe that he or she has no bias since a project proponent's partner is the surveyor's second cousin twice removed, but is well advised to make full disclosure of the connection before consideration of the project begins. There may in fact be no bias on the part of the surveyor/board member, but the mere possibility of bias may be enough to create an appearance of a conflict of interest.

Less obvious may be the surveyor's risk of a conflict of interest in his normal private practice. Consider the possibilities.

A surveyor agrees to make a survey for the owner of property whose neighbor just happens to be the surveyor's uncle. A concern arises when the survey seems to favor the neighbor/uncle, at which point the surveyor's client believes the surveyor's judgment was skewed as a result of the family connection. In the ensuing argument between the two warring neighbors, the surveyor has lost credibility with his client; he has an *appearance* of a conflict of interest. The courage required in this example must be exercised at the contracting stage when the surveyor and his client first discuss the work to be done. The surveyor should at least notify the potential client of the family connection, but with any possibility of dispute over the results of the anticipated survey, the prudent surveyor might best withdraw from the project. Such a decision may require courage in a time of scarce work.

Another example involves a surveyor providing services for adjoining housing projects. Project 2 will depend upon the completion of some of the roads in Project 1 for access to utilities. The developer of Project 1 decides to delay progress on those particular roads in order to delay next-door competition. All may be fair in love and war—and the housing industry—but to the extent that the surveyor is involved in the decision-making process of Developer 1, he may have a conflict of interest or at least an *appearance* of a conflict. It may be difficult to convince Developer 2 that the surveyor is innocent of complicity in the matter; after all, it was the surveyor's office that planned the connection of utilities between the two projects, etc., etc.

A third example is the topographical survey of a site on which a commercial use had long since been abandoned. At the rear of the property, the surveyor finds apparent remnants of an old dump: rims of steel barrels appearing at ground surface; mysterious discoloration on patches of exposed soil; bits and pieces of half-buried debris. The surveyor, working for the current owner of the land parcel, understands that his work is to be used by planners working for a prospective purchaser of the site. The surveyor understands that by defining the unwelcome evidence he is alerting the purchaser to a possible clean-up cost, news that the surveyor's client, the current owner, will not receive happily. So there is the surveyor's dilemma: to make a client angry by introducing uncertainty about the suitability of the property—and its worth—or shirking a social responsibility while assuming a possible liability to the new owner. To whom does the surveyor owe loyalty—the client or the future owner? And isn't the surveyor suddenly in a conflict of interest between the two parties? We might introduce the public interest as a third party in the dilemma, which only makes the conflict more complicated for the surveyor.

Surely, in this third example, any ethical surveyor would meet the professional, ethical, and social responsibility of detailing the suspicious evidence. Courage is often required to do the right thing; to deliver bad news to a client is the surveyor's lot in life. In each of these examples, the surveyor has found himself in a conflict of interest, and not necessarily due to the surveyor's own poor planning or lack of judgment. How one resolves the conflict is a matter of professionalism, integrity, and courage.

## II. Confidentiality

A question of the confidentiality of surveying service is often raised. The point is that the results of a surveyor's work are usually made public through the monumentation of corners and the recording of plans; but a client's interests may suggest that the surveyor *not* publish his work through monumentation or recording of plans. It becomes a matter of client confidentiality. But does the surveyor have a right to reuse the information acquired in the unrecorded survey? Who owns the information and results of a survey? To what extent should surveyors share their data with other surveyors? Is it ethical to use the results of a survey for client A in the process of a survey for client B?

Many of these questions may be dealt with in the wording of carefully scripted contracts. A survey contract should contain clauses specifying the delivery items as well as the ownership of the surveyor's work product. For instance:

> *Delivery Items.* Upon completion of the survey *Surveyor* shall furnish to the client the original and copies of the resulting plat or map and copies of *Surveyor's* boundary description and any formal reports prepared by *Surveyor.* The boundary description prepared from the survey embodies *Surveyor's* opinion of the location of the boundary lines of the site and is not to be construed as a certification to quality of title or location of title to the property.
>
> *Ownership of Plans and Documents.* All field notes, field data, soil boring logs, reports, deeds and title reports, calculations, working drawings, estimates and other documents acquired or prepared by *Surveyor* as instruments of service shall remain the property of *Surveyor.*

Attorneys have advised that these "instruments of service" rightfully belong to the surveyor and are his to use in appropriate circumstances, such as future surveys of adjoining or nearby properties. That is a legal principle that *might* be recognized in a litigated dispute, but the prudent step is for the surveyor to make it a matter of agreement in her signed contract with her client.

A more difficult question arises when a survey has been completed for a prospective purchaser of land. Perhaps that client intended to develop the property but wanted information on parcel area, topography, soil conditions, available public utilities, and so on. The client had a purchase-and-sale agreement that enabled him to hire a surveyor to acquire the information. Let's assume that the survey was accomplished under contract containing the above clauses, but that the client decides not to purchase the property—after spending several thousand dollars on the surveyor's fee. Assuming the client agreed to the contract clauses similar to those suggested here, we would suppose the surveyor is free to use the data acquired in his service under the contract.

On the other hand, the client may claim proprietary right in the information. In this writer's opinion, it is ethically clear that a duplicate of the topographical plan and the soil report that was delivered to his client are not available for sale to another prospective purchaser of the property. But what about all that information—instruments of service—in the surveyor's files? Presumably it would be easy enough to reconstruct the topo plan. The soil test results reside in notebooks in the surveyor's files and are available for reformatting in a new formal report. Ethically, the surveyor should be able to use this information in advising a second client but *without charging again* for the fieldwork required to acquire the data. (Most codes of ethics enjoin against double charging or reselling of a service.)

Whether the surveyor is willing to take advantage of a contractual right in this example is also a question of business judgment. If the surveyor ever hopes to work again for the earlier client, he would be wise not to just make the information available to the next customer. A better way would be to advise client B that the information is available but has been paid for by client A. Rather than requiring the surveyor to repeat the fieldwork, perhaps a deal between clients A and B could be worked out. Both clients should respect the surveyor for her good judgment in the matter.

The sharing of information between surveyors raises other issues. Such cooperation between competing surveyors may be both useful and professionally commendable. In the best of circumstances, surveyors recognize their positions in the community as peer relationships. They seek each other's advice and support as they do in membership in local surveyor associations. But when a surveyor opens his files to others in an effort to be helpful, he may be creating liability exposure for himself. One attorney pointed out that the surveyor's files, computations, and working drawings contain a record of the decision-making process in which are hidden all attempts at resolution, trial and error approaches, and perhaps even calculations found later to be in error. That attorney advised against opening surveyors' files to anyone—clients or other surveyors.

The accumulated site-specific knowledge belongs to the surveyor as a professional practitioner. It should be protected both through contracting practices and file management procedures.

# Possible Surveyor Liability Relating to Tree Encroachments

25

Richard F. Bales

Most real estate attorneys and title insurers, when discussing issues concerning encroachments as disclosed by a survey, think in terms of buildings, sheds, or fences. However, there is a whole body of law relative to encroachments of trees onto adjoining property. In today's urban and litigious society, it would not be uncommon for the practicing attorney to one day become embroiled in a neighborhood dispute concerning the encroachment of a tree onto or from a client's property. This article will summarize Illinois law concerning this subject.

## I. Boundary Tree Encroachments

There are three types of tree encroachments. The first is an encroachment of the tree itself onto the adjoining land. This was the issue in *Ridge v. Blaha.*[1] In *Ridge,* the court allowed the issuance of an injunction, restraining the defendant from destroying a tree located on the boundary line between two parcels of land. The court stated the general rule that a tree growing on such a boundary line is the common property of the adjoining owners as tenants in common. Thus, neither party would have the right to cut, injure, or destroy the tree

1. 520 N.E.2d 980 (Ill. App. Ct. 1988).

without the consent of the other.[2] This is the case, even if the majority of the trunk of a tree is located on one owner's land.[3]

The court also noted that the location of the tree is determined by the location of the trunk and not the roots or branches; specifically, the exact location of the tree vis-à-vis the adjoining land is determined at the point the tree emerges from the ground.[4] Thus, the fact that a tree's roots alone cross a boundary line is insufficient to create common ownership, even though the tree derives nourishment from both parcels of land. Thus, one who goes onto a neighbor's land in order to cut down a tree that straddles the boundary line between two properties could be guilty of trespass.[5]

For a case concerning the possible negligence of a municipality in allowing the continued existence of bushes and an evergreen at a street intersection, see *First National Bank in DeKalb v. City of Aurora.*[6]

## II. Encroachments of Roots or Branches

The second and third type of encroachment consists of tree roots or tree branches that encroach onto adjoining property. Some cases leave the landowner to the common law remedy of self-help in regards to encroaching vegetation from adjoining property.

For example, in *Merriam v. McConnell,*[7] the court refused to enjoin the defendant from growing box elder trees on his property, even though the plaintiff alleged that the box elder bugs emanating from the trees were a private nuisance. The court, commenting on a Massachusetts case, stated:

> The Court thought it wiser to adopt the common law practice of leaving the neighbor to his own protection if harm results to him from this exercise of another's right to use his property in a reasonable way, than to subject that other to the annoyance, and the public to the burden of actions

---

2. *Ridge* indicates, however, that an adjoining landowner could trim the branches of a "boundary tree" that hung over his property, as long as the trimming would not interfere with the use of the tree by the other landowner. See further discussion *infra; see also* 2 C.J.S., *Adjoining Landowners* § 55; 1 Am. Jur. 2d *Adjoining Landowners* § 22.

3. *See, e.g.,* Kimber v. Burns, 97 N.E. 671 (Ill. 1912) (in *Kimber,* a tree was 8 inches on the plaintiff's lot, and 22 inches on defendant's lot, yet the court held that the defendant wrongfully removed the tree).

4. *Ridge,* 520 N.E.2d at 982.

5. Kimber v. Burns, 97 N.E. 671; *see also* Simpson v. City of Gibson, 164 Ill. App. 147 (Ill. App. Ct. 1911).

6. 353 N.E.2d 309 (Ill. App. Ct. 1976). For a similar case, but one involving a utility company's alleged negligence, see *Hammond v. SBC Commc'ns,* 850 N.E.2d 265 (Ill. App. Ct. 2006). *See also* Jamison v. City of Zion, 834 N.E.2d 499 (Ill. App. Ct. 2005).

7. 175 N.E.2d 293 (Ill. App. Ct. 1961).

> at law, which would be likely to be innumerable and, in many instances, purely vexatious.[8]

The *Ridge* court also gave its seal of approval to self-help: "[I]t would not have been proper to restrain defendants from trimming overhanging branches *which cause them damage.*"[9] Note, though, that the court was silent as to whether self-help is allowed when the tree branches are not damaging the adjacent owner's land. However, the court did note with approval *Robinson v. Clapp,* which did not require damage as a condition precedent to self-help.[10] Furthermore, the court in *Simpson* stated that if a tree were located on private property, but the branches hung over the street, the city would have the right to remove the tree branches, without compensation to the adjoining owner, if the branches "damaged the street or were an inconvenience to the public in the use of the street."[11] On the other hand, while *Merriam,* as noted above, does not state outright that damages is a prerequisite for self-help, it does speak of one exercising self-help "if harm results to him."[12]

What appears to be dispositive of the issue is noted in 2 C.J.S. *Adjoining Landowners* sec. 54; it seems clear that the landowner has the right to remove tree encroachments whether they cause damage or not:

> Even though a landowner has sustained no injury by the intrusion on or over his land of the branches or roots of a tree or plant on adjoining land, he may cut off the offending branches at the boundary line.[13]

The court in *Bandy v. Bosie*[14] noted *Merriam* with approval. In *Bandy,* trees on the defendant's lot dropped sap and leaves on the plaintiff's property, and roots from these trees entered and damaged the plaintiff's sewer line, causing water to back up in his basement. The plaintiff sought injunctive relief and damages, claiming the trees were a nuisance. The court in *Bandy* refused to grant relief and damages, stating that:

> We do not consider trees that drop leaves on neighboring lands or trees that send out roots that migrate to neighboring lands and obstruct drainage to necessarily constitute a nuisance. . . . [U]nder the circumstances here, to permit the falling of leaves or the migration of the roots to give rise to injunctive relief would unduly promote litigation over relatively minor matters. Usually, the damage from the offending leaves would be

8. *Id.* at 295.

9. *Ridge,* 520 N.E.2d at 984 (emphasis added).

10. 32 A. 939, 942 (1895).

11. *Simpson,* 164 Ill. App. at 148.

12. *Merriam,* 175 N.E.2d at 295.

13. *See also* Bonde v. Bishop, 245 P.2d 617 (Cal. Ct. App. 1952); Cannon v. Dunn, 700 P.2d 502 (Ariz. 1985).

14. 477 N.E.2d 840 (Ill. App. Ct. 1985).

> minimal, and the accurate locating of the source of the offending roots would be difficult and expensive.[15]

Damage was not so minimal in *Mahurin v. Lockhart*,[16] where the court reached a different conclusion. Here, the plaintiff sought to recover damages for personal injuries sustained when a dead branch extending over his property fell and struck him. The court discussed the traditional rule of nonliability for dangerous conditions arising out of purely natural causes. This rule developed at a time when land was mostly unsettled and uncultivated. As the landowner was unable to remedy all of the dangerous conditions arising out of purely natural causes, he was thus shielded from liability out of necessity.

The court stated that while there may be reasons to continue this traditional rule in rural areas, there is little or no reason to apply it in urban or suburban areas, as it would not be unduly burdensome for the property owner to inspect his property and take reasonable precautions against dangerous natural conditions. Thus, the court held:

> [A] landowner in a residential or urban area has a duty to others outside of his land to exercise reasonable care to prevent an unreasonable risk of harm arising from defective or unsound trees on the premises, including trees of purely natural origin.[17]

Although an appellate case, *Chandler v. Larson*,[18] discussed and distinguished several of the previously mentioned cases, and is probably the leading Illinois case dealing with these types of encroachments. In *Chandler*, the court was faced with the issue of whether an urban landowner can state a cause of action for negligence where damage to his property results from the growth of roots of a tree which is located on the defendant's adjoining property. Here, the roots extensively damaged the plaintiff's garage. The court ruled that the defendant, an owner of urban property, owed adjoining landowners a duty of reasonable care and that the plaintiff's amended complaint stated a legitimate cause of action for negligence.

The court distinguished *Merriam v. McConnell* by noting that in *Merriam*, the plaintiff did not allege negligent conduct on the part of the defendant. Quoting *Merriam*, the *Chandler* noted:

---

15. *Id.* at 842. But see *Olson v. Westerberg*, 119 N.E.2d 413 (Ill. App. Ct. 1954) for a contrary result (abstract only); *see also* 2 C.J.S. *Adjoining Landowners* § 54.

16. 390 N.E.2d 523 (Ill. App. Ct. 1979).

17. 390 N.E.2d at 524–25. Note that the court in *Bandy* cited *Mahurin* with approval, stating that "[t]he *Mahurin* decision appears to be in line with a trend to place greater responsibility upon the owner of the lot where the tree is located." 477 N.E.2d at 841. The court went on, though, to find in favor of the defendant, stating that *Mahurin* was distinguishable from *Bandy* in that in *Bandy*, there was no allegation that the trees constituted a danger or that they were negligently maintained.

18. 500 N.E.2d 584 (Ill. App. Ct. 1986).

> The law is that defendants themselves would have to be guilty of some carelessness, negligence or willfulness in bringing, or helping to bring, about a harmful condition in order to entitle plaintiff to the relief sought in this particular prayer. The complaint does not allege that they were guilty in any of these ways.[19]

The court in *Chandler* also stated that "[i]n the present case, we have two adjoining owners of city lots, and consequently, the *Merriam* decision would not be directly applicable."[20]

The court also discussed *Bandy,* commenting that this case involved parties who owned adjoining city lots. In this respect, the case was similar to *Chandler.* However, the court distinguished the two, noting that the plaintiff in *Chandler,* unlike the plaintiff in *Bandy,* made an allegation of negligence on the part of the defendant.

The *Chandler* court cited with approval the holding in *Mahurin,* noting that in both cases the plaintiffs and defendants owned adjoining city lots.[21] Also, the plaintiffs in both cases stated causes of action for negligence in their complaints. Consequently, the court in *Chandler* commented:

> The *Mahurin* opinion is in agreement with the trend to place greater responsibility upon the owner of the property where the tree is located. . . . [T]he present matter is most analogous to the factual situation in *Mahurin v. Lockhart.*[22]

## III. The Restatement (Second) of Torts

The traditional rule concerning liability of landowners to third parties for harm caused by natural conditions on the land is set forth in section 363 of the RESTATEMENT (SECOND) OF TORTS:

> (1) Except as stated in Subsection (2), neither a possessor of land, nor a vendor, lessor, or other transferor, is liable for physical harm caused to others outside of the land by a natural condition of the land.

19. *Chandler,* 500 N.E.2d at 587.

20. 500 N.E.2d at 588. As noted above, *Merriam* appeared to involve adjoining landowners living in a wooded, suburban setting.

21. Note, though, that the holding in *Mahurin* was not predicated on the fact that the lots were in an urban setting; indeed, the *Mahurin* court stated just the opposite: "While there are many reasons to continue the traditional rule [of nonliability] in regions that are largely rural, there is little or no reason to apply it in urban and other developed areas. . . . [We] hold that a landowner *in a residential or urban area* has a duty to others outside of his land to exercise reasonable care. . . . 390 N.E.2d at 524–25 (emphasis added).

22. 500 N.E.2d at 587–88.

(2) A possessor of land in an urban area is subject to liability to persons using a public highway for physical harm resulting from his failure to exercise reasonable care to prevent an unreasonable risk of harm arising from the condition of trees on the land near the highway.

Comment (a) to subsection (1) states that this is the rule, even though the landowner may be aware of the dangerous natural condition and the expense necessary to remedy the condition is slight. On the other hand, comment (e) to subsection (2) makes it clear that the urban landowner has a duty to travelers to inspect his property and take reasonable precautions against dangerous conditions. However, the court in *Mahurin* attacked this traditional rule, stating that the greater duty owed to persons using the highways should also be extended to adjoining landowners in urban and residential areas.

Note that subsection (1) of section 363 of the RESTATEMENT (SECOND) OF TORTS refers to *natural* conditions of the land. Subsection (1) of section 840 of this RESTATEMENT states that, generally speaking "a possessor of land is not liable to persons outside the land for a nuisance resulting solely from a natural condition of the land." Section 840, comment (a), states that vegetation is deemed "natural" when it is "not in any way the result of human activity." On the other hand, section 839 imposes a duty upon an adjoining landowner to abate the condition when the encroaching vegetation is "artificial" and subjects the adjoining landowner to liability for failure to abate the "artificial condition."[23]

To date, Illinois has not limited this RESTATEMENT rule that distinguishes "artificial" and "natural" conditions.[24] Note, though, that the court in *Melnick v. C.S.X. Corporation* discusses this rule at length. It states that while a few states follow this rule, most states do not. It notes further that some courts have commented that the distinction between "artificial" and "natural" vegetation is unworkable and should not control the issue of a landowner's duty to an adjoining landowner.[25]

## IV. Practical Application of Case Law

Assume, then, that a client comes into your office and tells you that the branches or roots of a neighbor's tree are encroaching into his yard. The trunk of the tree is solely in the neighbor's yard. What do you tell him?

23. An example of an "artificial condition," as noted in illustration 4 to comment (a) to section 840, would be eucalyptus trees planted by an adjoining landowner near his boundary line; *see also* Noel, *Nuisances from Land in its Natural Condition,* 56 HARV. L. REV. 772 (1943).

24. Nichols v. Sitko, 510 N.E.2d 971 (Ill. App. Ct. 1987); Burns v. Addison Golf Club, Inc., 514 N.E.2d 68 (Ill. App. Ct. 1987); Eckburg v. Presbytery of Blackhawk, 918 N.E.2d 1184 (Ill. App. Ct. 2009).

25. 540 A.2d 1133, 1136–37 (1988); *see also* Eckburg v. Presbytery of Blackhawk, 918 N.E.2d 1184 (Ill. App. Ct. 2009); Ortiz v. Jesus People USA, 939 N.E.2d 555 (Ill. App. Ct. 2010); Belton v. Forest Preserve Dist. of Cook Cnty., 943 N.E. 2d 221 (Ill. App. Ct. 2011).

The preceding analysis of Illinois case law indicates that he may cut off the branches or roots, but only up to the property line. He cannot cross into the neighbor's yard in order to cut off the branches or roots. Note that the doctrine of self-help would appear to be an appropriate measure in Illinois, even if the roots or branches are not causing damage to the client's property.

What if the client tells you that he fears that if he removes the roots of the offending tree, the tree will die? May he still remove the roots?

There does not appear to be any Illinois case on point in this area. However, note the Oklahoma case, *Higdon v. Henderson.*[26] There, the plaintiff sought damages for the destruction of a tree located on the boundary line between plaintiff's and defendant's property. Defendant, while constructing a residence on his lot, excavated land solely on his property. The excavation cut open and exposed the roots of the tree, which eventually died. The Supreme Court of Oklahoma acknowledged the same rule as set forth in *Ridge v. Blaha, supra,* namely, that a tree located on the boundary line between two adjoining landowners is the common property of both owners and that neither owner has the right to damage or destroy the tree without the consent or permission of the other party. However, the court further stated that this rule is tempered by the fact that an abutting owner has the right to use his property in a reasonable way and, conversely, not in an unreasonable way. The court stated that the defendant was excavating on his own lot, and nothing more, which was not an unreasonable use of his property. Thus, the court denied relief, stating that the resultant injury to the tree did not create a right to recover damages.[27]

Thus, it would at first appear that one could remove the roots with impunity. This was not so in *Holmberg v. Bergin,*[28] a Minnesota case where the defendant planted a tree on his property, which eventually grew to a point where the roots were damaging the plaintiff's property. The Minnesota Supreme Court let stand the decision of the trial court, stating that an injunction ordering the complete removal of the tree, rather than just the roots, was warranted, since the removal of only the roots would weaken the tree's stability, thereby endangering the homes of both parties in the event of a storm or high wind. In light of this decision, it would seem appropriate for the attorney to advise the client to not remove the roots, as the client could be liable in the event the tree were to fall and injure people or property. Instead, the attorney might consider asking for an injunction, ordering the removal of the entire tree. But as for tree removal, see the Wrongful Tree Cutting Act.[29] This act provides for treble damages in the event a party cuts a tree "which he did not have the full legal right to cut or caused to be cut."[30]

26. 304 P.2d 1001 (Okla. 1956).

27. *Id.* at 1002.

28. 172 N.W.2d 739 (Minn. 1969).

29. 740 Ill. Comp. Stat. Ann. 185/1 et seq.

30. 740 Ill. Comp. Stat. Ann. 185/2. *See also* 2 C.J.S. *Adjoining Landowners* § 56; 1 Am. Jur. 2d *Adjoining Landowners* § 23.

If the roots or branches have damaged the client's property, the attorney may wish to file suit for damages. In that event, it would appear, from the preceding analysis of existing case law on the subject, that the suit should be one in negligence and not in nuisance. As the court in *Merriam* stated:

> [P]laintiffs have recovered damages . . . only where a human agency has intervened in a negligent, careless or willful way to turn [natural forces] into a nuisance. . . . In order to create a legal nuisance, the act of man must have contributed to its existence. Ill results, however extensive or serious, that flow from natural causes, cannot become a nuisance. . . . [A] nuisance cannot arise from the neglect of one to remove that which exists or arises from purely natural causes. But when the result is traceable to artificial causes, or where the hand of man has, in any essential measure, contributed thereto, the person committing the wrongful act cannot excuse himself from liability. . . . [31]

## V. Possible Defenses

By now, the question may have arisen: can the defendant/adjoining landowner assert any defenses to a possible cause of action? For example, assume the client told his neighbor that he was going to cut off the branch that encroached into his yard. The attorney for the neighbor fires off a letter to the client, asserting that the neighbor has acquired that space occupied by the branch by virtue of adverse possession. Is such a statement tenable?

Most likely not. Note that it is true that a landowner owns at least as much of the space above the ground that he can occupy or use in conjunction with the land.[32] However, in order to establish title via adverse possession, there must be, among other things, some unequivocal act of ownership upon the land.[33] A tree branch extending over a neighbor's land—even for at least twenty years, the requisite statutory period[34]—does not appear to qualify as an "unequivocal act of ownership."[35]

At some point in all this, the attorney may ask himself: "Is the title insurance company or surveyor liable for not showing the tree branch or root encroachment on, respectively, the title insurance policy or survey of my client?" Accordingly, discussion follows.

---

31. Merriam v. McConnell, 175 N.E.2d 293, 295–96 (Ill. App. Ct. 1961) (citing 1 Wood, Nuisances 148–49 (3d ed. 1893)).

32. United States v. Causby, 328 U.S. 256 (1946).

33. Brooks v. Bruyn, 24 Ill. 373 (1860); Joiner v. Janssen, 421 N.E.2d 170 (Ill. 1981).

34. 735 Ill. Comp. Stat. 5/13–101 (1992) (formerly, Ill. Rev. Stat. 1991, ch. 110, sec. 13–101).

35. See, e.g., *Dempsey v. Burns,* 118 N.E. 193 (Ill. 1917), wherein the court said that the enclosure of a small lot by two smooth wires was not a sufficient act of possession. *See also* Burlew v. City of Lake Forest, 433 N.E.2d 353 (Ill. App. Ct. 1982).

## VI. Possible Title Insurance Company Liability

Title company liability for any possible encroachment problem is, for the most part, predicated on the title company giving "extended coverage" over those general exceptions relating to survey matters.[36] There are five general exceptions; the two generally considered to relate to survey matters and unrecorded easements are:

> (2) Encroachments, overlaps, boundary line disputes, or other matters that would be disclosed by an accurate survey and inspection of the premises.
>
> (3) Easements, or claims of easements, not shown by the public records. Assuming that coverage over these exceptions was given, consider first the encroachment of the adjoining landowner's tree branch onto the client's land.

The title company would probably deny any claim tendered by its insured relative to the encroachment of a tree branch or aboveground tree root onto its insured's property. As has previously been discussed, the insured has had its own remedy from the beginning—one of self-help. This would be the case, regardless of whether the branch had caused damage to the insured's property. As the Supreme Judicial Court of Massachusetts stated: "His remedy is in his own hands."[37]

Similarly, the title company would probably be able to deny liability based on policy coverage. The ALTA 2006 Owner's Policy contains several "Exclusions from Coverage." Exclusion 3(a) is [d]efects, liens, encumbrances, adverse claims, or other matters created, suffered, assumed, or agreed to by the insured claimant." Any insured owner who acquires the land would do so with full knowledge of any encroaching tree branch. Any such owner who chose not to avail himself of the remedy of self-help would quite likely be estopped from asserting a claim.[38]

Finally, it is quite possible that the tree encroachment is of such a de minimis nature so as not to warrant the response of the title company. *Brownstone Condominium Association v. Geller*[39] was concerned with the alleged encroachment of nine bolts into a neighbor's building. The court stated:

36. Attorneys Guide to Title Insurance 4–42 et seq. (Ill. Inst. for CLE, 1980).

37. Michalson v. Nutting, 175 N.E. 490, 491 (Mass. 1931).

38. See, for example, *Malkin v. Realty Title Ins. Co., Inc.*, 223 A.2d 155 (Md. 1966), wherein the court noted this exclusion in denying relief to the plaintiffs. In this case, the title policy inadvertently failed to include an exception relative to the existence of a paved roadway along the Westerly 33.8 feet of the insured land. The court affirmed the judgment of the lower court, noting that the plaintiffs had inspected the property and had reviewed a survey that disclosed the roadway; the plaintiffs thus had independent knowledge of the fact that the roadway burdened their property.

39. 415 N.E.2d 20 (Ill. App. Ct. 1980).

> [T]he mere fact there is a trespass does not warrant any affirmative action by the Court. The close quarters of an urban society demand flexibility when we encounter an intrusion on our personal bubble. In a metropolitan area of seven million people, none can deny that their rights, real or imagined, are infringed upon with regularity. The limitations on the judicial system make it obvious that each of these transgressions cannot be litigated.[40]

An encroachment of underground tree roots would probably elicit a similar response from the title company. The reason for this is that it appears that such an encroachment would most likely fall outside the boundaries of title insurance coverage.

General exception (2) concerns those matters that would be disclosed by a survey or an inspection of the land. A survey or inspection would not reveal the location of underground tree roots. Thus, the fact that this exception was deleted from a title insurance policy insuring land burdened by underground tree roots would not result in title company liability.

General exception (3) deals with unrecorded easements or claims of easements. *Kayfirst Corp. v. Washington Terminal Co.*[41] dealt with the 7-foot encroachment of an underground step footing onto adjoining property. The court stated that the step footing, located 4 feet below the surface of the land, was not an implied easement by necessity or an easement by prescription since the use of adjoining land was not apparent. Thus, the fact that the title policy issued to the plaintiff was subject to the aforementioned general exception (3) did not relieve the title company from liability because of the encroachment.

The court acknowledged that the step footing was an encroachment. However, it also noted that it was 4 feet beneath the surface of the ground. Since it was not visible from the surface, it would not have been disclosed by a survey or inspection of the land. Thus, similarly, the fact that the title policy was also subject to general exception (2) did not exempt the title company from liability.

The court went on to discuss the insuring provisions of the title policy. It noted that the title company insures against loss or damage which the insured may sustain by reason of four defects in the title of the insured. Specifically, it noted insuring provisions (2) and (4):

> (2) Any defect in or lien or encumbrance on such title;
>
> (4) Unmarketability of such title.

The court concluded that while there was no liability on the part of the title company because the policy was subject to general exceptions (2) and (3), it did hold that the encroachment of the underground step footing was an encumbrance

40. *Id.* at 22.
41. 813 F. Supp. 67 (D.D.C. 1993).

on the title; it also stated that the encroachment rendered the plaintiff's title unmarketable as well:

> It has long been established that an encroachment may encumber title or render title unmarketable, particularly if the encroachment is substantial or interferes with the free development of the subject parcel.[42]

At first one might maintain that an underground step footing is analogous to underground tree roots, thus resulting in title company liability in the event of an encroachment thereof. However, the court in *Kayfirst* made its determination based on the encroachment's size and the fact that the slab footing interfered with the commercial development of the site. It is doubtful that a court would ever conclude that tree root encroachments would be "substantial" or that they could interfere with the development of the adjoining land. This is especially so when one remembers the decision in *Higdon v. Henderson, supra,* wherein the court allowed the excavation of the defendant's land, even though the removal of tree roots caused the destruction of the tree.

## VII. Possible Surveyor Liability

The current version of the Minimum Standard Detail Requirements for ALTA/ACSM Land Title Surveys were adopted by the American Land Title Association (ALTA) and the American Congress on Surveying and Mapping (ACSM) in 2011.[43] While these national survey standards are designed to be used by surveyors when surveying any type of property, they are usually used only when surveying commercial, industrial, or larger vacant and residential properties. If a surveyor prepared a land title survey pursuant to these survey standards and failed to show an encroaching tree branch or aboveground tree root on his plat of survey, would he be liable to his client?

Part 5 of these standards is entitled "Fieldwork," and it details what information should be shown on the plat of survey. Section 5(C) is captioned "Lines of Possession, and Improvements along the Boundaries." Relative thereto, the surveyor is advised that his fieldwork shall include the following:

> 5(C)(i): The character and location of evidence of possession or occupation along the perimeter of the surveyed property, both by the occupants of the surveyed property and by adjoiners, observed in the process of conducting the survey.

42. 813 F. Supp. at 76.

43. Richard F. Bales & Gary R. Kent, *The 2011 ALTA/ACSM Land Title Survey Standards,* 25(4) Probate & Property 12 (2011).

> 5(C)(ii): The character and location of all walls, buildings, fences, and other improvements within five feet of each side of the boundary lines, observed in the process of conducting the survey.
>
> 5(C)(iii): Without expressing a legal opinion as to the ownership or nature of the potential encroachment, the evidence, location and extent of potentially encroaching *structural* appurtenances and projections observed in the process of conducting the survey, such as fire escapes, bay windows, windows and doors that open out, flue pipes, stoops, eaves, cornices, areaways, steps, trim, etc., by or onto adjoining property, or onto rights of way, easements or setback lines disclosed in Record Documents provided to the surveyor. (emphasis added)

Paragraph 5(C)(i) requires the surveyor to show "[t]he character and location of evidence of possession or occupation along the perimeter of the surveyed property" on his plat of survey. However, in light of the previous discussion of adverse possession, any argument based on surveyor liability vis-à-vis this paragraph hardly seems tenable. Furthermore, and more important, in light of paragraph 5(C)(iii), which is described below, this section does not really concern encroachments.

Pursuant to paragraph 5(C)(ii) of these survey standards, the surveyor has to disclose "[t]he character and location of all walls, buildings, fences, and other improvements" on his plat of survey. Clearly, a tree, or any part thereof, does not constitute an improvement.

Paragraph 5(C)(iii) makes it clear that a surveyor's duty under the standards is to disclose only *structural* encroachments on his plat of survey. Because a tree branch or tree root is clearly not something of a "structural" nature, it does not appear that a surveyor would be liable for failing to show an encroaching tree branch or tree root on his plat of survey.[44]

However, surveyors in Illinois rarely use these land title survey standards when performing surveys of 1–4 residential properties. While most of these surveys are not "certified" to a specific standard, these surveys are usually performed to Illinois boundary survey standards.[45]

Section 1270.56(b)(6)(M) of these boundary standards requires the surveyor to disclose *visible* "physical evidence of possession or occupation either way from the exterior lines of the survey." Section 1270.56(b)(6)(N) of the standards requires the surveyor to show *visible* "evidence of improvements, rights-of-way, easements, or use" on the plat. (emphasis added) Clearly, the surveyor would not be liable for failing to show on his plat any encroaching underground tree roots.

---

44. See *Rozny v. Marnul,* 250 N.E.2d 656 (Ill. 1969) for what has been described as the "seminal" case dealing with the liability of surveyors to third parties. *See also* Richard F. Bales, *Land Surveyor Liability to Third Parties in Illinois,* 95(3) Ill. B.J. 136 (2007).

45. Ill. Admin. Code tit. 68, §§ 1270.56 et seq.

The land title survey standards require the surveyor to disclose only those encroachments that are *structural* in nature. These boundary survey standards contain no such limitation. In fact, these standards do not even include the word "encroachment." Instead, as noted above, the surveyor must disclose only evidence of "possession, occupation, and improvements." It appears, therefore, that the surveyor has no duty to show any encroaching tree branches or roots on his boundary plat of survey.

There are only a handful of Illinois cases that deal with tree encroachments. Nonetheless, it is evident that these cases contain a myriad of issues. It is possible that the general practitioner will never have to deal with a client who has a tree encroachment problem. However, in the event that he does, it is hoped that this article will provide some insight.

# Early English and American Surveying[1]

# 26

Mitchell G. Williams

## I. Introduction

The mapping of boundary lines of real property is as ancient as the emergence of civilization in Egypt and Mesopotamia four thousand years ago. In Egypt, property lines of fields along the Nile were measured and marked with physical monuments so that the owners would be able to locate their lands after the annual flooding of the Nile subsided. The Code of Hammurabi, as well as the Hebrew scriptures, prescribed harsh penalties for those who moved or destroyed property markers. There are surviving copies of surveys of Roman tenements and towns. There are some primitive maps from England drawn in the Middle Ages, and there is a survey map dating from 1300 that refers to a settlement of a boundary dispute in 1150. In the sixteenth century, large-scale mapping by cartographers and scientists using compasses for measuring angles and directions and wooden rods or metal chains to measure distances, as well as sophisticated techniques, such as the method of "intersection" or triangulation (by which a point on the ground could be located by measuring the angle at each end of a fixed measured baseline to that point and using simple trigonometry) became common practice.

At the same time, the mapping of private property holdings in England developed into its own specialty, which came to be referred

1. Portions of this article appeared in 41(1) SURVEYING AND MAPPING 77 (1981).

to as land surveying, and in England and on the continent, surveying further divided into other specialized branches: geodetic surveying, cadastral surveying, military surveying, and construction surveying. Until recently, the instruments and measurement techniques of what we now know as "land surveying" were not as sophisticated or precise as those used in what came to be known as geodetic surveying. The relatively short length of the parcel boundaries in land surveying did not require the surveyor to take into account the curvature of the earth or the convergence of the meridians. From the sixteenth century until the mid-nineteenth century, a compass and chain sufficed to perform most survey work, including the large-scale laying out of the sections and townships for the U.S. Public Land Survey. By the end of the nineteenth century, chains had been replaced by steel tapes and the compass had been replaced by the surveyor's transit and theodolite, both of which had telescopic sights and could measure angles to the nearest minute of arc. By the end of the twentieth century, for measuring distances of more than a couple hundred feet, steel tapes were replaced by electronic distance meters using lasers. Today, Global Positioning System (GPS) receivers are in common use to survey even moderately sized properties to a high degree of precision.

The early development in England of land surveying has left its mark in numerous ways on real property law in the United States and on the system of public land established by the United States government in 1785 to govern the location, disposition, and settlement of the federal territories. Even a cursory overview of the history of land surveying can help explain certain archaic concepts and rules of law, as well as why two surveyors today, with surveying techniques capable of a precision of a few millimeters, may still disagree on the "correct" location of a boundary line.

## II. Early Surveying in England and America

Printed in 1523, Here Begyneth a right Frutefull Matter: and hath to name the boke Surveynge and Improvementes by John Fitzherbert is the earliest known English text on land surveying. At the time, a "surveyor" (derived from a French term meaning to "look over") was another name for the steward of a manor. The steward's responsibility was to produce a written report on the nature and quality of the crops and other products produced by the tenants of the landlord. Each tenant typically had a narrow strip of land on the manor designated for his use, and the tenants as a group had the use of the "commons": areas of the manor that were farmed or used for pasturage collectively by the tenants for their own needs under the "custom" of the manor. Beginning in the fifteenth century and continuing into the seventeenth century, due to economic pressures, landlords had begun to consolidate the tenants' strips into larger fields enclosed by hedges or walls. They also "enclosed" all or parts of the commons for the landowner's use and benefit. Since enclosure caused the tenants to lose their traditional holdings, enclosure was resisted, sometimes violently. With the acceleration of enclosure during the sixteenth century, a new profession arose: surveyors who could measure and lay out the newly enclosed fields and boundary lines, and make maps and written descriptions of the new parcels. Surveyors would also restore the original property boundaries and corner markers of the enclosed areas which had been lost or obliterated by disgruntled tenants from those maps and written descriptions. During this same period, property surveys were also being done in the American colonies to lay out towns and cities as well as to document property boundaries.

There were eleven surveying texts printed in English in the sixteenth century, and seven in the seventeenth. The most influential in the United States was John Love's GEODAESIA: OR THE ART OF SURVEYING AND MEASURING OF LAND MADE EASIE, which was first printed in England in 1688. John Love was a surveyor in North Carolina and Jamaica before writing the book. The twelfth and thirteenth editions were published in the United States in 1793 and 1796.

## III. Surveying Practice

The basic instruments used by surveyors through the end of the seventeenth century were the surveyor's compass (often called a circumferentor), an instrument for measuring horizontal angles called a "theodolitus" or a "topographical instrument" depending on the author of the text describing it, and a chain for measuring distances. The compass and theodolitus were usually set up on a staff or a tripod. The surveying compass was a magnetic compass with a needle surrounded by a horizontal ring which was divided into 360 degrees, and frequently each degree was divided into 6 parts, each representing 10 minutes of angle. The ring around the compass was equipped with two vertical sighting bars 180 degrees apart. The surveyor would align the sights with the point to be located, and then read the bearing of that point from the location of the compass by looking at the compass needle. A surveying compass from the late nineteenth/early twentieth century is shown in figure 1.

**Figure 26.1** An example of a late nineteenth/early twentieth century surveying compass. On this instrument, the two vertical sights are mounted on a sighting bar, which is attached to a post mounted on the rim of the compass. This arrangement allows for taking a bearing to a point that is higher or lower than the compass.

The theodolitus had a circular plate marked with degrees and an alidade, or sighting bar, that pivoted at the center of the instrument. The alidade could be used to measure the angle between two different points. The theodolitus frequently had a compass or a compass disk as well. In the seventeenth century, there were examples of early theodolites that had the alidade mounted on a stand above the horizontal plate so that it could also be used to measure vertical angles. The true theodolite did not come into common use for property surveying until the nineteenth century, although it was used for geodetic surveying and for large-scale mapping in the eighteenth century. A modern example of a form of theodolite commonly known as a surveyor's transit is shown in figure 2.

**Figure 26.2** An example of a twentieth century surveyor's transit. Telescopic sights were not common on land surveying instruments used for property line surveys until the nineteenth century.

The basic instrument for measuring distances was the Gunter's chain, which was literally a chain made up of iron links. Edmund Gunter was an English astronomer who "invented" the chain. It is 66 feet long (or 4 rods, the statutory rod in England being 16.5 feet[2]) and consists of 100 links, each 7.92 inches long. Vestiges of these units remain in older legal descriptions and also in roads that were laid out in the nineteenth and early twentieth century: a significant number of those roads are 33 or 66 feet wide. One of the reasons the chain was widely adopted is that it is divided into 100 links: measurements could be recorded in decimal form rather than feet and inches, and that made calculations far simpler than dealing with inches and fractions of inches. It also facilitated calculations of area: one reason Gunter chose 66 feet is that 10 square chains would equal 1 acre.

There are two relatively early examples of large-scale surveys of private property boundaries that are comparatively well documented. The first is the Down

2. The terms pole and rod were common units of measure in England, and there were local variations as to how long they were: 18 feet was not uncommon. In 1227, Edward I fixed the length of a "statutory rod" at 16.5 feet. See *When Is a Rod Not 16.5 Feet? (More Times than Not)* by Knud E. Hermansen, chapter 27 in this volume, for a more detailed discussion.

Survey, done in the late seventeenth century in Ireland by English surveyors. The second is the initial U.S. Public Land Surveys, the first of which were done at the end of the eighteenth century. Both were done with the instruments and techniques discussed above (although by the early nineteenth century, American surveyors were using surveying transits and solar compasses instead of surveying compasses). Both are examples of "original" surveys, which were performed to establish the boundary lines of land parcels, as opposed to "retracement" surveys, which are done to reproduce the originally established boundary lines in their original location. An examination of the Down Survey and the early U.S. Public Land Surveys will help to explain why modern surveyors sometimes disagree about the correct location of the lines described in a deed description.

## IV. The Down Survey

Ireland rebelled against the English occupation in 1641, and the fighting continued until 1652 when the last of the rebels surrendered. The English Parliament financed its suppression of the rebellion in part by soliciting funds from investors, who would be repaid at the conclusion of hostilities in the form of land in Ireland forfeited by the rebels. In 1652, Parliament passed the Act of Settlement, which specified in detail the class of Irish landowners whose property was forfeit, outlined provisions for appeal, and included the soldiers of the occupying army in the settlement by dividing the forfeited lands evenly between the investors and soldiers. Ten counties were set aside for the satisfaction of the investors' and soldiers' claims, while the remaining forfeited lands in the rest of the country went to the English Crown.

In order to allocate the forfeited lands to the investors and soldiers, it was necessary to know which lands were forfeit, where they were located, their size and dimensions, and the nature of the land. This obviously required a land survey of the counties involved, and Parliament authorized one, which came to be known as the Down Survey. The making of the survey was first put in the hands of Benjamin Worsley, the Surveyor-General, and he prepared a list of the forfeited lands, barony by barony, while the inhabitants were being physically displaced across the Shannon River. Worsley's proposal for the survey, however, had a number of shortcomings: the lands to be given to the investors and soldiers would be allocated before their location, description, and nature had been determined, opening the way for bitter and long boundary disputes and creating the possibility that the quality and value of the land given to a settler would depend on when he staked his claim, with latecomers shortchanged.

These defects were brought to the government's attention by Sir William Petty, who had been serving as Chief Physician to the Parliamentary Army at the close of the rebellion. He outlined his own plan for a survey and argued persuasively enough that he was given the contract for the survey in December 1654, with the expected delivery date only one year later.

Petty was a minor figure of the Enlightenment. Born in 1623 in England, he went to sea as a youth, studied medicine on the continent, and became a professor of anatomy as well as of music. His education was broad, and he left a large number of essays, notes, and books on numerous subjects. He has been credited with founding the science of statistics, and his major work, POLITICAL ARITHMETIC (published in 1690), is concerned with the "algebra" of government and economics. His work is filled with elaborate calculations and formulae: he deduced that the optimum allowance was 3 acres of good land per capita and computed that, on the basis of available statistics, 10 million Englishmen were worth 13 to 14 million Frenchmen. His rational arithmetic approach was characteristic of the Enlightenment. Thomas Jefferson had a representative sample of his works in his library.

Petty's instructions to his surveyors were in two parts: (1) the nature of the information required and (2) technical advice on how to acquire it. His primary goal was to have a detailed topographic map of Ireland, showing all ditches, hedges, hills, fields, forests, houses, towns, etc., with the boundaries of all of the forfeited baronies superimposed. The surveyors were also to note the profitability of the land and the quality of the soil. The job of apportioning the land to be given to each soldier and investor could be done completely on paper, including the specification of the boundary features of the parcels conveyed, so that there would be no grounds for dispute and no necessity for resurveys of individual parcels.

The technical instructions are interesting for the light they shed on seventeenth-century surveying practice. All angles were to be measured with the circumferentor, or surveying compass, and distances were to be measured with the chain; the boundaries were to be located by intersection (triangulation) from a number of points so that a check could be maintained on the work. The common lines of each barony were to be run out independently by two surveyors to control their location. The surveys were to be referenced to the "true," rather than the magnetic, meridian, and a "tying line" would be run from a conspicuous landmark to "some remarkable angle or other marker in each surround [boundary]," so that the boundaries could not be lost with the passage of time. The fieldwork was to be plotted on graph paper (an innovation) at a scale of 40 perches [rods] to the inch, and any error of closure was to be found by examining the courses and distances as plotted, and the courses and distances would be adjusted to minimize the error. There were no mathematical checks on closure. Areas were computed by the simple expedient of counting the number of grid boxes within each parcel and multiplying it by the area of the grid box. A plat (or map) of each barony was to be made at a scale of 80 perches to the inch, and all the plats and field books, each checked by two sworn surveyors, were to be deposited in a public registry in Dublin.

Petty delivered the complete survey in the year promised. He had had a staff of approximately one thousand people, and the work was systematically divided so that the fieldwork, plotting, and title searching were all being done simultaneously. The survey was, relatively speaking, accurate. Petty published an atlas of Ireland based on the survey, and that atlas was not improved upon until the large-scale ordinance mapping of the eighteenth century.

## V. The Early U.S. Public Land Surveys

At the end of the American Revolution, the new states had ceded their claims to land west of the Allegheny Mountains to the new government. In 1784, the Continental Congress appointed a committee, chaired by Thomas Jefferson, to prepare a plan for the sale of the cession lands. Although the land to be mapped and distributed was the result of the forced dispossession of the Native Americans by right of conquest rather than as punishment for rebellion, the problem facing Congress was similar to that faced by the English Parliament in 1652, including disagreement about whether the cession lands should be divided at all prior to settlement. The southern states, led by Virginia, wanted the land opened to immediate settlement and claim, with the surveys and maps to be made after the fact, while the New England bloc favored a rational division before settlement and sale.

The Virginia delegates argued that the settlement would be delayed by prior survey and, perhaps more important, that a prior survey would lock settlers into parcels of land that might not agree too well with the natural lay of the land. The reasons for the New England position were summarized in a letter to George Washington from William Grayson, the chairman of the survey committee:

1. There certainly must be a difference in the value of the land in the different parts of the country, and this difference cannot be ascertained without an actual survey.
2. Because the eastern states, where the lands are more equally divided than in any other part of the continent, were generally settled in that manner.
3. The idea of a township, with prospects of support for religion and education, would be an inducement for neighbors of the same religious sentiments to confederate for the purpose of settling together.
4. The southern method would defeat this end by introducing the idea of indiscriminate locations which would have a tendency to destroy all these inducements to emigration.
5. The exemption from controversy on account of boundaries for all time.
6. The right to form governments for themselves would induce emigrants from all parts of the world and insure a settlement of the country in the most rapid manner.

7. The expense and delay would be too great to divide the territory into fractional parts by actual survey.
8. The method of laying out the same into squares is attended with the least possible expense, there being only two sides of a square to run in almost all cases.
9. It superseded the necessity of courts for the determination of disputes.
10. It excludes all formalities of warrants, entries, locations, returns and caveats, as the first and last process is a deed.[3]

All of these arguments in favor of a rational, but not necessarily rectangular, subdivision before settlement and a deed registry were familiar to most educated Americans and Englishmen of the period. In fact, they had been developed at length in the writings of Sir William Petty in the latter half of the seventeenth century.

## VI. Political Arithmetic, Jefferson, and the Rectangular System

In his later writings, Petty had become enamored of an "arithmetic" of society. His practical concern for assuring that the value of the land distributed to the English settlers in Ireland be equal was transformed into a formula of 3 acres per capita and schemes to give every man 40 acres. It was the algebraic Petty that Thomas Jefferson found most congenial. Jefferson himself proposed to give 50 acres of land to every free man in Virginia, and he had a lifelong obsession with perfect regularity in both the layout and operation of his farms. He wanted to parcel out the cession lands into "balanced" tiers of states. Jefferson's fascination with regular geometry, like Petty's, may be attributed to his surveying background. Petty was very proud of his Down Survey and Jefferson of his father's maps of Virginia. But in Jefferson's case, Petty himself was a decisive influence. In 1769, Jefferson bought Petty's SURVEY OF IRELAND, and he read with pleasure his POLITICAL ARITHMETIC. Petty's influence can be seen in Jefferson's original plans for the survey of the cession lands. It is not surprising that he would turn to the Down Survey as a model. There had been no successor to the Down Survey. It stood alone as the only large-scale systematic boundary survey until the U.S. Public Lands survey got underway.

The primary difference between Jefferson's proposal in his 1784 committee report and the Down Survey was the rectangular subdivision of the territories into units 10 geographical miles square, with each unit further subdivided into parcels 1 mile square. Otherwise, the survey was to be done by the "needle," i.e., by compass, but corrected for variation from true north; all corners were

3. LOWELL STEWART, PUBLIC LAND SURVEYS 3 (1976).

to be marked and noted, the character of all land recorded, and the location of watercourses, mountains, and all "remarkable and permanent things" plotted. Similar to the Down Survey's public registry in Dublin, a land register was to be established where all the plats and field books were to be stored.

The square subdivision of the land was a Petty-inspired idea, one which he could not utilize in the Down Survey because the country was already settled, and only forfeited lands were being subdivided, not the entire landscape. The specification of geographic miles and squares reveals that Jefferson was slightly out of his element and that he was dealing in the kind of abstraction that characterized the POLITICAL ARITHMETIC. The two basic problems with Jefferson's scheme were the use of the so-called geographical mile, the length of which changes with latitude (Jefferson had to fix it at 6086.4 feet), and the contradictory stipulations that the towns be squares and their east and west boundaries be true meridians. The first of these was resolved in 1785 by the specification of statute miles as the basic unit of length, but the incompatibility of squares and meridian lines was not resolved until the Act of 1800.

Other than detail changes in the sizes of the sections and townships, the Rectangular System was instituted as Jefferson outlined it, although he had left the committee by 1785 and had nothing more to do with it directly. The Act of 1800 resolved the problem of the meridian convergence and, more important, also established the principle that the corners, as set by the surveyors would control over their theoretical location, which, of course, has resulted in a system of rectangular subdivision which is hardly ever rectangular, although it is systematic. This result was abetted by a 1786 ruling allowing the magnetic, rather than the true, meridian to be used in running lines. Technically, the execution of both surveys by compass and chain was identical, although mathematics replaced Petty's graphical methods for determining closure and area in the U.S. Public Land Surveys.

## VII. Conclusion

The precision of early land surveys, including the Down Survey and the early U.S. Public Land Surveys, was not high, and, as a result, the legal descriptions derived from them could be difficult to reliably locate on the ground. If the description did not form a mathematically closed figure, and if the error of closure was large, another surveyor could locate the boundaries as described and be feet or more away from their original location. The larger the parcel surveyed, the greater the error in location.

There were many reasons for this lack of precision The chain used by the original surveyor may not have been exactly 66 feet long due to errors in manufacture, the length of the chain itself may have changed through use as the links bent and were pulled further apart (resulting in a measurement shorter than the actual distance between the points), surveyors may have failed to pull the chain tight enough to eliminate the sag (resulting in a measurement longer than the actual distance between the points), and surveyors may have failed to measure the distance in a horizontal plane. Also, any distance less than 1 link, or 7.92 inches, would have to be estimated. The measurement of angles was more susceptible to error. The use of the surveying compass resulted in errors in the bearings of property lines: magnetic north is not the same as true (or geographic) north, and the difference between them (known as magnetic variation) varies from place to place, and also over time, because the magnetic north pole moves. The presence of metal or iron near the compass would also cause the needle to point away from north. Although a theodolitus had a scale marked in intervals of 10 minutes of arc, it is unlikely that angles could be read to that precision: there was no effective method to level the instrument, there were no telescopic sights, and the technology used to engrave the 10-minute intervals itself was prone to error.[4]

There were also errors in transforming the survey results into a map of that survey. Even with tables of trigonometric functions and logarithms, calculations were difficult, time-consuming, and prone to errors of arithmetic, and had to be done by hand. The calculation of the error of closure of the survey was sometimes done by plotting the courses and distances on paper and measuring the gap, and sometimes by mathematical calculation. There were no statistical methods to adjust the error of closure, only rules of thumb. Even after the method of "least squares" had been developed by Carl Frederic Gauss in the late eighteenth century and first published in 1809, the calculations were too complex for most land surveyors and rarely used in property surveying. Surveyors in this period were not, in general, mathematically sophisticated. Calculations of area, which was one of the most important numbers a landowner would want to know, were particularly difficult unless the property was rectangular or could be easily divided into rectangles and right triangles. Many surveyors used crude formulae, for example dividing the length of the perimeter of the property by four and squaring the result. That works for a square, but underestimates the area of a circle by 21 percent and overestimates the area of a rectangle that is three times as long as it is wide by 33 percent. Area calculations on most surveys were little more than very rough estimates.

There is a pattern in the above discussion of errors. English and American law established rules on the priority of evidence in the event of discrepancies in legal descriptions. A common formulation is:

1. Natural monuments (rivers, lakes) or artificial monuments (stakes, pipes) if specifically set forth in the description;
2. Specific references to a boundary in another deed or map of record;
3. Recitals of distances along boundary lines;

4. See *When Is a Rod Not 16.5 Feet? (More Times than Not)* by Knud E. Hermansen, chapter 27 in this volume for a table showing the extent of error resulting from errors in angle measurements.

4. Recitals of bearings of, or angles between, boundary lines; and
5. Recitals of area.

The order reflects the increasing likelihood of errors in the bearings, angles, distances, and area in the original survey and description.

Most of the surveys done by property surveyors today are "retracement surveys," which require determining the location of the original property lines that were laid out when the parcel was created. The older the date of the original survey, the more likely it is that two surveyors will come to different opinions about the correct location of the boundaries. If the differences are significant, the parties should enter into a boundary line agreement, if feasible, to fix the location of the disputed line, and record it in the public records. If an agreement cannot be reached, the solution may be litigation, and a court will determine which survey more accurately reflects the *original* survey lines (not the lines shown on the last survey that was made, but the ones established when the parcel was created). Once that determination is made, the description of record can be amended, and, due to the relatively high precision of today's surveying equipment and techniques, any future retracement survey of the corrected description is unlikely to be controversial.

## Further Reading

Love, John, Geodaesia (1688).
Petty, William, The History of Surveying of Ireland (1851).
Petty, William, The Petty Papers (1927).
Price, Edward T., Dividing the Land (1995).
Richeson, A.W., English Land Measuring to 1800 (1966).
Stewart, Lowell, Public Land Surveys (1976).
Wills, Garry, Inventing America (1979).

# 27

# When Is a Rod Not 16.5 Feet? (More Times than Not)[1]

Knud E. Hermansen[2]
PhD, PLS, PE, Esq.

The science of geometry and mathematics is exact. The infinite depths of stellar space are measured with such exact nicety that the position of stars and planets can be calculated to the fraction of a second of time. . . . How can it be that in the ascertainment of one line of so small an area, bounded by four lines only, a difference of from 8 to 24 feet arises? It is evident that the methods pursued, and not a defective science, have brought about the different results, different maps. *Warren v. Boggs,* 332, 111 S.E. 331 (W. Va. 1922)

As experienced title attorneys and paralegals know, measurements along the same boundary vary between old and new surveys. At first impression, there does not seem to be a logical reason for the sometimes large disparity between the measurements found in the deed and the modern surveyor's measurements. Consequently, litigators and the surveyor's client have the impression that the surveyor is at worst incompetent or at best negligent in surveying the historical boundaries and adding or taking away land. The fact is that most times, differences in the measurements do not reflect any change in boundary location or the addition or loss of land. Variations between old and new measurements are in fact common and should raise questions only if there were no differences.[3] Nevertheless, clients and

1. An edited version of this article appeared in 6(5) PROBATE & PROPERTY 8 (1992).
2. Knud Hermansen has a PhD in civil engineering from Pennsylvania State University and a JD from West Virginia University. He is a professor at the University of Maine and operates a consulting firm offering surveying, engineering, and legal services.
3. W. Mining & Mfg. Co. v. Peytona Cannel Coal Co., 8 W. Va. 406, 431 (1875).

litigation involving property boundaries frequently require a rational explanation to help explain the difference between the measurements cited in the records and more recent measurements.

> The science of mathematics is exact, but the different results reached in its application by different surveyors, is sometimes startling to the layman, when applied to what appears to be an ordinary survey.[4]

> The original surveys of lands in the older States of the American Union were exceedingly deficient in precision. This arose from two principal causes; the small value of land at the period of these surveys, and the want of skill in the surveyors. The effect at the present day is frequent dissatisfaction and litigation. Lots sometimes contain more acres than they were sold for, and sometimes less. Lines which are straight in the deed, and on the map, are found to be crooked on the ground. The recorded surveys of two adjoining farms often make one overlap the other, or leave a gore between them. The most difficult and delicate duty of the land-surveyor, is to run out the old boundary lines. . . .[5]

To comprehend the basis for the difference, knowledge of the surveyor's duty and some historical information is required. The surveyor's duty in regard to surveying historical boundaries is often described as "following in the footsteps of the original surveyor."[6] Unfortunately, searching for footsteps involves searching for recollections, markings, monuments, and records that typically range in age from 50 to 300 years old. The intervening time has taken its toll on this evidence through decay, fire, flooding, construction, unintentional destruction, deceit, ignorance, and the unavailability or incompetency of reliable witnesses, to name a few.[7]

> [B]ut old surveys are not to be so tested. Most perfect in the beginning they are constantly undergoing change and decay, until by wind, fire, rottenness, and the acts and frauds of men, their evidences lie only in memory and hearsay.[8]

---

4. Zirkle v. Three Forks Coal Co., 103 W. Va. 614, 626, 138 S.E. 371 (1927), quoted in Warren v. Boggs, 90 W. Va. 330 (1922).

5. W.M. Gillespie, A Treatise on Land-Surveying: Comprising the Theory Developed from Five Elementary Principles; and The Practice with the Chain Alone, the Compass, the Transit, the Theodolite, the Plane Table, etc. (1881).

6. Rivers v. Lozeau, 539 So. 2d 1147, 1151 (Fla. 1989). While the concept has always been applied, the words that so aptly describe the surveyor's charge are said to have first appeared in a speech titled "The Judicial Functions of Surveyors," by Chief Justice Cooley of the Michigan Supreme Court, read before the Michigan Association of Engineers and Surveyors.

7. Ulman v. Clark, 100 F. 180, 187 (D. W. Va. 1900); Northumberland Coal Co. v. Clement, 95 Pa. 126 (1880); Kennedy v. Lubold, 88 Pa. 246 (1878); Ralston v. Groff, 55 Pa. 276 (1867); Cutts v. King, 5 Me. 482, 487 (1829).

8. Kennedy v. Lubold, 88 Pa. at 246.

> Monuments referred to in deeds are often perishable; as trees, wooden buildings, or fences; or slight and temporary; as a stake, or a stake and a few loose stones, intended to be supplied by something of a more permanent character. They serve to point out at the time, to the parties in interest, the bounds of the land conveyed. After these monuments are gone, and such a period of time has elapsed, that no one can be found who remembers to have seen them, or can testify as to their location; uniform continued occupancy, by buildings, fences or other equivalent indications of ownership is evidence that the land was located according to the original monuments. These monuments perish; and time sweeps away those who could point out where they stood. . . .[9]

To further compound the problem, preventative or curative actions were impeded through ignorance, denial, or the seemingly prohibitive costs associated with surveying. As a result, the deed descriptions so often copied from one conveyance to the next are seldom as reliable or unpretentious as reliant parties would hope. The following is a brief explanation for some of the many errors and inaccuracies in older measurements.

## I. Equipment Precision

The equipment used during the early surveys was not as refined or precise as modern survey equipment.[10] The typical equipment used in early land surveys consisted of a compass and chain. In some rural areas, this equipment continued to be employed up into the 1960s.[11] The typical compass and chain[12] was seldom able to obtain measurements better than the nearest 1/4 degree (15 minutes) in direction and nearest link (7.92 inches) in distance.[13]

---

9. Cutts v. King, 5 Me. at 487.

10. Winding Gulf Colliery Co. v. Campbell, 72 W. Va. 449, 467–68 (1913).

11. In fact, it would not be unusual to see this method employed at the present time for some large, rural woodland parcels.

12. "The ordinary surveyor's chain is sixty-six feet, or four poles long, composed of one hundred links, each connected to the other by two rings, and furnished with tally marks at the end of every ten links." W & L.E. Gurley, A Manual of the Principal Instruments Used in American Engineering and Surveying 141 (1878).

13. A "finer cut" was impractical since traverse tables were generally limited to the nearest 15 minutes. Gillespie, *supra* note 5; Gurley, *supra* note 12.

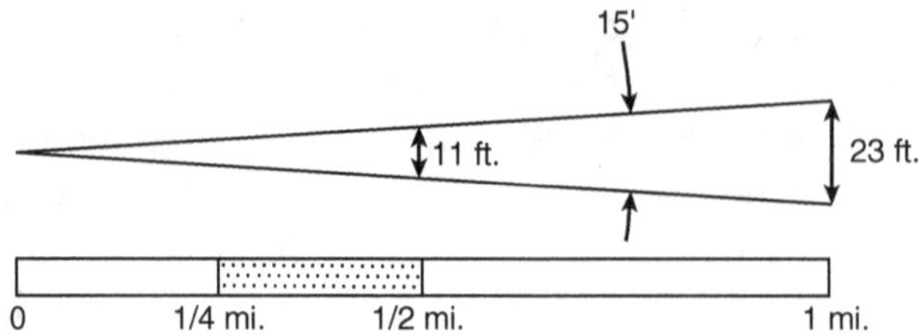


| Distance | | | | | | |
|---|---|---|---|---|---|---|
| Angular | 16,500 ft. | 5,280 ft. | 1,650 ft. | 1,000 ft. | 500 ft. | 165 ft. |
| Uncertainty | (1000 rods) | | (100 rods) | | | (10 rods) |
| **10°** | 2887.1 | 923.9 | 288.7 | 175.0 | 87.5 | 28.9 |
| **1°** | 288.0 | 92.2 | 28.8 | **17.5**** | 8.7 | 2.9 |
| **30'** | 144.0 | 46.1 | 14.4 | 8.7 | 4.4 | 1.4 |
| **15'** | 72.0 | **23.0*** | 7.2 | 4.4 | 2.2 | 0.7 |
| **1'** | 4.8 | 1.5 | 0.5 | 0.3 | 0.1 | 0.0 |
| **30"** | 2.4 | 0.8 | 0.2 | 0.1 | 0.1 | 0.0 |
| **15"** | 1.2 | 0.4 | 0.1 | 0.1 | 0.0 | 0.0 |
| **1"** | 0.1 | 0.0 | 0.0 | 0.0 | 0.0 | 0.0 |

**Figure 27.1** The error caused by a 15-minute deviation in direction is shown by the figure. A 15-minute deviation in direction results in an error of 23 feet per mile.* Similarly, a 1-degree deviation in 1,000 feet results in an error of 17.5 feet.**

The typical compass did not have magnification and only a rudimentary method to measure the slope (if at all). The limitations of the compass were well known among the early surveyors and members of the Bar.[14] The magnetized needle frequently lost its magnetism or was subject to changes in the magnetic pole or variances caused by electric storms, the Aurora Borealis, and nearby magnetic attractions (local attractions).[15] In some cases, metal shavings or impurities were found to reside in the brass compass housing that drew the needle off along certain directions.[16]

The chain, the other piece of ancient survey equipment, was heavy and unwieldy. It was difficult to suspend without introducing considerable sag. Links soon stretched, became bent, clogged with debris, or kinked, adding to the uncertainty of measurements.[17]

14. Lodge v. Barnett, 46 Pa. 477 (1864); Hagey v. Detweiler, 35 Pa. 409 (1860); Ralston v. Groff, 55 Pa. 276 (1867); Blasdell v. Bissell, 6 Pa. 258 (1847).

15. Report of the Commissioner on the Variations of the Magnetic Needle, Variations of the Magnetic Needle, State of Maine (1866) [hereinafter Report of the Commissioner].

16. Cox v. Couch, 8 Pa. 147 (1848); Gurley, *supra* note 12.

17. *Lodge*; Heaton v. Hodges, 14 Me. 66 (1836); Gurley, *supra* note 12. "If a chain's long links are held together by three rings, which was common enough, then there are eight wearing surfaces per link or 800 wearing surfaces per chain. If each surface wore 0.01 inch, the chain would be eight inches

The adoption of the vernier transit and much lighter steel tape by many surveyors in the late 1800s and early 1900s allowed practitioners to measure directions to the nearest minute and distances to the nearest 1/100th of a foot, every 100 feet. (Compare this to modern equipment, which can consistently measure angles to the nearest second and a distance (as far as visibility permits) to the nearest 1/100th of a foot. Using the newest equipment—satellite receivers—visibility between stations is no longer a factor.)

## II. Practitioners

The training and skill of some past practitioners left much to be desired.[18] Rigorous training and formal education for surveyors were haphazard or nonexistent. One or more surveyors seemed to practice in every locale where their only attributes seem to have been a sense of direction, hemp rope, or consistent pace, and a passable talent to draw lines. Their practice was questionable and would amount to fraud by today's standards.[19] Licensing, which was intended to remove the charlatans, was not mandatory in many states until the later half of the 1900s.[20] Even after licensing of surveyors, many licensing requirements did not require a test or proof of skills before issuing a license to practice.

Assuming the surveyor had the minimum skill and knowledge, the help the surveyor employed seldom did.[21] The surveyor arriving at the site with a trained or semi-trained field crew was almost unheard of in the past. Help was more often than not the client and men hired from the local population. A survey crew in the early days was supervised by the surveyor or a trusted deputy who generally operated the compass or transit. The remainder of the survey crew (on a large survey) consisted of two chainmen hired from among the local population, two or more axe men to cut and mark line, a cook, and a cook's helper to clean utensils and help pack supplies. Training of the chainmen was rudimentary at best and left much to be desired in the resulting accuracy of the distances.[22]

> [I]t was not error for the court to call the attention of the jury to the fact that defendant's measurements were made by a "baker attended by a

longer." Patrick Tascano, *Gunter's Chain*, 51(3) SURVEYING & LAND INFORMATION SYSTEMS 155 (1991).

18. Many practitioners will candidly admit that the early surveyors in George Washington's time were of the highest caliber. The skill and knowledge of the average surveyor subsequently declined. The trend appeared to reverse at some point midway in the twentieth century. *See, e.g.*, Mahon v. Duncan, 13 Pa. 459 (1850).

19. Blain v. Woods, 145 W. Va. 297, 306, 115 S.E.2d 88 (1960).

20. The first licensing act was attributed to Wyoming in 1907. L.C. BISHIP, SURVEYING IN WYOMING DURING TERRITORIAL DAYS AND NOW (1957).

21. Ralston v. Groff, 55 Pa. 276 (1867); Cox v. Couch, 8 Pa. 147 (1848); *Blasdell v. Bissell*, 6 Pa. 258 (1847).

22. Reilly v. Mountain Coal Co., 54 A. 29 (Pa. 1903); Omenstetter v. Kemper, 6 Pa. Super. 309 (1898); Fisher v. Kaufman, 33 A. 137 (Pa. 1895).

> tinsmith under the supervision of a lawyer." This is not such departure from judicial gravity as to call for a reversal.[23]

## III. Terrain and Site Conditions

Present day practitioners and landowners sometimes fail to remember what the terrain and site conditions were like at the time of the early surveys. Virgin timber several feet in diameter, both standing and fallen, presented formidable obstacles to thwart the surveyor in measuring a straight line through the forest.[24] Hostile Indians,[25] foreign powers seeking control of the wilderness, squatters not interested in paper title, wild animals, disease, and lack of shelter and nutritious food took their toll. Under the circumstances, surveyors were more concerned with their surroundings and well-being than their measurements.

> The difficulty of making an accurate survey by courses and distances, under the conditions obtaining in that country at the time this survey was made, were very great. It was a rough heavily timbered country, making it hard to see between stations, distant from each other, and slow and irksome to chain directly from station to station; but it was comparatively easy to select accessible points for corners, and practically guess at the courses and distances. To this we must add the circumstances that there was then a mad rush of speculators into this region for land at two cents an acre, and consequent pressure upon the surveyors, well calculated to induce resort to the easiest and quickest method of achieving results.[26]
>
> [I]n the wilderness in which those early surveys were made, it was practically impossible to avoid mistakes.[27]

Even after the virgin timber was removed and the land settled, the surveyor's ability to measure accurately was hampered by dense growth brought on by the now abundant sunlight and rich soil on what had once been shaded forest floor. Blazes once made to mark the boundaries were lost when the timber was removed or decayed. The present twenty-minute drive to the courthouse took a day or more in the past on roads that were mere muddy paths or covered with snow or debris. As a result, records were not always obtained, and the previous measurements for the property and measurements for the adjoining property were not always compared before recording a new description or map.

23. Omenstetter v. Kemper, 6 Pa. Super. 309, 313 (1898).
24. Gwynn v. Schwartz, 32 W. Va. 487, 492–93 (1889).
25. Ulman v. Clark, 100 F. 180, 183 (D. W. Va. 1900).
26. State v. King, 64 W. Va. 546, 579–80 (1908).
27. Winding Gulf Colliery Co. v. Campbell, 72 W. Va. 449, 471 (1913).

## IV. Land Values

Many attorneys continue to use the same description written a hundred years ago. This practice not only fails to uncover latent problems but also ignores the law of economics. The same parcel worth several hundred thousand dollars today was frequently purchased for pennies when the last survey was performed.[28] In the past, the cost of having the land surveyed may have been more then the price to purchase the land. Under these conditions, speed was more important than fastidious measurements.[29] The carelessness that caused the omission or overlap of a few acres at 10 cents an acre was not worth the 25 cents required to resurvey and correct the error. The landowner purchasing 400 acres was not concerned with overlaps or a deficiency of a few acres.[30] Needless to say, a deviation of a rod or two on a measurement would not have caused any concern whatsoever.

## V. Procedures

The procedures employed by early surveyors leave much to be desired by today's standards. Old survey texts are filled with suggestions that were generally unknown or ignored by the early survey practitioner.[31] Surveyors were cautioned that frequent use of the chain would inevitably cause the links to stretch and eventually require the surveyor to remove a link or two. The surveyor willing to achieve measurements accurate to a few feet was advised to avoid measurements using the compass at certain times since the compass needle tended to vary by a few minutes during these periods of the day.[32] Deviations caused by the shift in magnetic north over time and location were ignored even though the error amounted to several degrees in some cases.[33] Instructions packaged with new compasses were quick to warn the surveyor to hold the chain away from the compass, periodically sharpen and adjust the spindle, and relieve the static electricity that built up in the glass.[34]

> When, however, the glass becomes electric, the fluid may be removed by breathing upon it, or touching different parts of its surface with the

28. *King*, 546, 579–80 (1908); Simmons Creek Coal Co. v. Doran, 142 U.S. 417, 432 (1891).
29. *King*, 64 W.Va. at 579–80 (1908); Ralston v. Groff, 55 Pa. 276 (1867).
30. Collins v. Barclay, 7 Pa. 67 (1847).
31. *See, e.g.*, George L. Hosmer & Charles B. Breed, The Principles and Practice of Surveying (1st ed. 1906); Gillespie, *supra* note 5; Gurley, *supra* note 12.
32. "[O]wing to the influence of the sun, which, in summer, will cause the need to vary from ten to fifteen minutes in a few hours, when exposed to its fullest influence." Gurley, *supra* note 12, at 57. The diurnal change for Eastport, Maine, was found to average around 15 minutes. Report of the Commissioner *supra* note 15, at 17.
33. Hagey v. Detweiler, 35 Pa. 409 (1860).
34. Gurley, *supra* note 12.

moistened finger. An ignorance of this apparently trifling matter has caused many errors and perplexities in the practice of the inexperienced surveyor.[35]

Problems were so prevalent and generally ignored during surveys conducted in the early and mid-1800s that legislation was passed in many states requiring surveyors to periodically check their chain against a known line and note the deviation of their compass from a known meridian. Even the otherwise cautious surveyor was sometimes unaware of problems caused by iron ore deposits or other localized attractions sufficient to pull the needle off during a reading.

Ignorance of proper procedures or the speed necessary to survey large tracts in a short time resulted in paper surveys (i.e., protracted lines)[36] or surveyors pacing, using stadia[37] or slope chaining rather than making time-consuming horizontal measurements.[38]

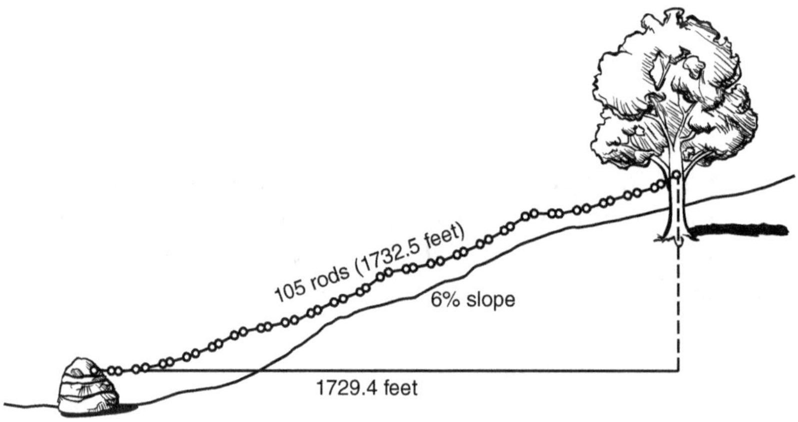


**Figure 27.2** The error resulting from measuring on the slope rather than a horizontal distance is shown in the above figure. At a 6 percent slope and distance of 105 rods, a 3-foot error will occur. The steeper the slope or the longer the distance, the greater the error.

In some cases, distances were estimated and directions approximated.[39] In other cases, haphazard corrections such as adding "one rod to each score"

35. *Id.*

36. West Virginia Pulp & Paper Co. v. Dodrill, 221 F. 780, 785 (N.D. W. Va. 1915); Ruffner's Heirs v. Hill, 31 W. Va. 428, 432 (1888); Packer v. Schrader Mining & Mfg. Co., 97 Pa. 379 (1881); Fisher v. Kaufman, 33 A. 137 (Pa. 1895).

37. Keta Gas & Oil Co. v. Jents, 110 A.2d 369 (Pa. 1955).

38. "[A]ll of the measurements were made in slope feet rather than horizontal feet. . . ." Vandetta v. Yanero, 157 W. Va. 220, 222, 200 S.E.2d 674 (1973); *Keta Gas & Oil Co.,* 110 A.2d 369; Cox v. Couch, 8 Pa. 147 (1848); Blasdell v. Bissell, 6 Pa. 258 (1847).

39. State v. King, 64 W. Va. 546, 579–80 (1908); Fisher v. Kaufman, 33 A. 137 (Pa. 1895).

for slope measurements were applied in an attempt to compensate for crude practices.[40]

> [I]t appeared, that at the time this survey was made, an excess of ten or twelve per cent had been allowed by the surveyors in other parts of the lines of said township. . . . [41]
>
> But the experience of the Courts has shown, that excess of admeasurement is so uniformly indicated in surveys of that early period, the Court is not prepared to say, that the excess, which was proved in this case, was evidence, which would warrant the jury in drawing an inference of fraud.[42]

The practice of actually running the boundary rather than traversing around the property forced many early surveyors to measure across obstacles or estimate the breadth of the obstacle rather than go around it. As a result, estimations were frequent. At other times, chains were laid on top of obstacles, or the chains curved around the obstacle rather than measuring the straight line distance between them.

Area, which is a product of the direction and distances, can be no better than the worst measurement. As a result, the area which is frequently of most concern to the layman is subject to the widest variations and exaggeration.[43]

> The acre of that day, as is and was well known, in the locations made in this State, was larger than the exact acre.[44]

## VI. Blunders

In the past, just as today, surveyors were prone to make mistakes. Early cases document many blunders that were discovered sometime after the survey.[45] It was not uncommon for the surveyor to lose his tally (the count of the number of chain lengths), transpose numbers, deviate from a straight line, misread the compass and chain, or make a miscalculation.[46]

---

40. Patrick Tascano, *Gunter's Chain,* 51(3) SURVEYING & LAND INFORMATION SYSTEMS 158 (1991); Dunn v. Hodges, 21 Me. 76 (1842); Otis v. Moulton, 20 Me. 205 (1841); Machias v. Whitney, 16 Me. 343 (1839); Heaton v. Hodges, 14 Me. 66 (1836).

41. *Heaton,* 14 Me. at 67.

42. *Machias,* 16 Me. at 348.

43. W. Mining & Mfg. Co. v. Peytona Cannel Coal Co., 8 W. Va. 406, 437 (1875).

44. Bussey v. Grant, 20 Me. 281, 286 (1841).

45. Day v. Wood Lumber Co., 78 W. Va. 19, 22 (1916); Holston v. Vaughan, 74 W. Va. 558, 560, 82 S.E. 390 (1914); Harman v. Alt, 71 S.E. 709 (W. Va. 1911); Stewart v. Doak Bros., 58 W. Va. 172, 175–76 (1905); Ulman v. Clark, 100 F. 180, 189 (D. W. Va. 1900); Gwynn v. Schwartz, 32 W. Va. 487, 495 (1889); Ruffner's Heirs v. Hill, 31 W. Va. 428, 437 (1888); W. Mining & Mfg. Co., 8 W. Va. at 418; *Machias,* 16 Me. 343; *Heaton,* 14 Me. 66.

46. Winding Gulf Colliery Co. v. Campbell, 72 W. Va. 449, 467–68 (1913); Ralston v. Groff, 55 Pa. 276 (1867); Lodge v. Barnett, 46 Pa. 477 (1864).

> [O]ld surveys were often inaccurate; and mistakes often made, in copying their descriptions into the patents; leaving out lines and putting north for south, and east for west; and in copying those descriptions into subsequent conveyances. . . .[47]

In some ways, errors were more likely to occur in the past than today. The literacy of the population in the early days led to many errors traceable to poor grammar, lack of formal education, and spelling.[48]

> The descriptions in deeds are usually prepared by surveyors who compose the calls with reference to the lines as they exist on the ground. Surveyors are not informed of or concerned with the fastidious refinement in the use of language favored in some courts.[49]
>
> Remoteness, land values, habits and education of the people, and other things, did not tend to promote accuracy.[50]

Other errors were a product of the time. Many of today's practitioners will no doubt attest to the fact that the invention of the typewriter was a welcome invention and prevented numerous errors previously caused by smudges, faded ink, and interpreting poor handwriting. The pencil and paper taken for granted by the modern practitioner and used to record information and jog the memory were rare and quite valuable in the past. The ink bottle and quill pen used by the early practitioners was not easily used in the field. The early surveyor was attuned to using knots on a thong, notches on wood, or sticks in a pouch to keep track of measurements. The slide rule and calculator, which have eased the burden of tedious calculations and removed the cause of many math errors, were beyond comprehension at the time most surveys were performed. All calculations were done by longhand.

This article is a brief summary of the many sources of errors in old measurements. A particular locale or name of an early surveyor may offer more particular reasons for differences. The attorney or paralegal, no less than the surveyor, should keep these facts in mind, especially when interpreting descriptions where directions are stated to the nearest degree or fraction of a degree and distances to the nearest rod or fraction of a rod.

In closing this report, it may not be improper to call attention to the fact that the various litigations and disputes about boundaries, which our courts of justice are constantly called upon to decide, are most of them either directly or indirectly the result of the present loose and imperfect method of conducting land surveys. This evil is not, however, it must be acknowledged, confined exclusively to the

47. *Winding Gulf Colliery Co.*, 72 W. Va. at 467–68.

48. MacCorkle v. City of Charleston, 105 W. Va. 395, 402, 142 S.E. 841 (1928); State v. Hicks, 76 W. Va. 508, 510–11 (1915); Wing v. Wood, 13 Me. 111 (1836).

49. *MacCorkle,* 105 W. Va. at 402, 142 S.E. 841.

50. State v. Hicks, 76 W. Va. 508, 510–11 (1915).

surveyors. Many of our lawyers, who are entrusted with the drafting of instruments of conveyance, are often deficient in the knowledge requisite to render their descriptions of land correct and to place them beyond the possibility of a misconstruction.[51]

51. Report of the Commissioner, *supra* note 15, at 74.

# Index

**G**

**Y**



**Z**